Orders of Magnitude

A History of the NACA and NASA, 1915-1990

by Roger E. Bilstein

The NASA History Series

National Aeronautics and Space Administration
Office of Management
Scientific and Technical Information Division
Washington, DC   1989

Table of Contents

FOREWORD

This is the third edition of *Orders of Magnitude*, a concise history of the National Advisory Committee for Aeronautics (NACA) and its successor agency, the National Aeronautics and Space Administration (NASA). At a time when American pride has been restored by the return of the Space Shuttle to flight,this edition reminds us of our first departures from the surface of the Earth and commemorates the 75th anniversary of the creation of the NACA--our first national institution for the advance of powered human flight. In less than half a century America progressed from the sandy hills of Kitty Hawk along the Atlantic Ocean into the vast "new ocean" of space. The pace of technological change necessary for such voyages has been so rapid, especially in the last quarter century, that it is easy to forget the extent to which aeronautical research and development--whether in propulsion, structures, materials, or control systems--have provided the fundamental basis for efficient and reliable civil and military flight capabilities. Thus it is fitting that this edition of NASA's *Orders of Magnitude* not only updates the historical record, but restores aeronautics to its due place in the history of the agency and of mankind's most fascinating and continuing voyage.

Perspective comes with the passage of time. Events since the last edition of *Orders of Magnitude* (1980) suggest that this nation's ability to sustain the enthusiasm and the commitment of public resources necessary for a vigorous national space program can, like the phases of our nearest celestial neighbor,wax and wane. The Apollo-Saturn vehicle that carried the first humans to the Moon was lofted not only by a remarkable mobilization of engineering research and know-how, but by the political will of a nation startled by the Soviet Union's display of space technology with *Sputnik* I, launched 4 October 1957.Universities and industry joined their considerable talents with NASA's to carry out the Apollo program's epoch-making exploratory missions in a truly national effort.

But responsiveness to changing national concerns is a hallmark of democratic government and the United States› preoccupations shifted to more earthbound concerns even before the Apollo program was completed. Public concerns such as energy resources, the environment, «guns and butter,» and fiscal restraint grew as a maturing aerospace technology broadened NASA›s mission as well as its rationale. Developed as a more economical approach to routine space travel than «throw away» boosters, the Space Transportation System with its reusable Shuttle orbiter was only one of NASA›s post-Apollo programs that reflected the new political climate of the 1970s and early 1980s.

As we approach the 1990s, however, my sense is that the nation›s interest in space exploration and exploitation, following a roller coaster of public interest (apathy, competing priorities, a brief moment in the sun with the pride of recovery) is on the brink of a period of such excitement, discovery,and wonder as to make the Apollo period pale in comparison. The scheduled voyages to Venus and Jupiter, the launch of the Great Observatories like the Hubble Space Telescope, the establishment of a permanent human presence in space with the space station Freedom, the development of a takeoff-to-orbit aircraft (the National Aerospace Plane), and the beginnings of engineering solutions to the technological requirements for expanding a human presence further into the solar system portend an era in which America, and indeed the world, will be bombarded with knowledge about the universe through which we pass so fleetingly. That knowledge, garnered in the finest traditions of intellectual endeavor that have characterized the history of the NACA and NASA,will foster a new vitality that will raise to new heights the cyclical pattern of public support for a strong national civil aeronautics and space program. While most of us are caught up in the changing events of each passing day, history--as this new edition of *Orders of Magnitude: A History of the NACA and NASA, 1915-1990* attests--reminds us of the continuities amid change and of our debt to those who have brought us the capability to write the next chapter in the history of humans out of Earth's bounds.

H. Hollister Cantus
Associate Administrator for External Relations
National Aeronautics and Space Administration
February 1987-November 1988

Preface

In 1965, Eugene M. Emme, historian for the National Aeronautics and Space Administration (NASA), wrote a brief survey of the agency entitled *Historical Sketch of NASA* (EP-29). It served its purpose as a succinct overview useful for Federal personnel, new NASA employees, and inquiries from the general public. Because people were so curious about the nascent space program, the text emphasized astronautics. By 1976, a revision was in order, undertaken by Frank W. Anderson, Jr., publications manager of the NASA History Office. With a different title, *Orders of Magnitude: A History of NACA and NASA, 1915-1980* (SP-4403), the new version gave more attention to NASA's predecessor, the National Advisory Committee for Aeronautics (NACA), although astronautics was still accorded the lion's share of the text. After a second printing, Anderson prepared a revised version, published in 1980, which carried the NASA story up to the threshold of Space Shuttle launches. Anderson retired from NASA in 1980.

As NASA approached the 75th anniversary of its NACA origins in 1915, a further updating of *Orders of Magnitude* seemed in order. In addition to its original audience, the book had been useful as a quick reference and as ancillary reading in various history courses; Anderson's graceful, lucid style appealed to many readers, including myself. The opportunity to prepare a revised survey was an honor for me.

Anderson›s original discussions of astronautics have remained essentially intact; these are represented in the concluding section («Enter Astronautics») in chapter 3 and by chapters 4, 5, 6, and 7 in this latest version. In recognition of the NACA›s acknowledged contributions to aeronautical progress, I wrote the first three chapters, carrying the story up to the origins of NASA in 1958. Although chapters 4 through 7 are basically unchanged, I have included a more detailed summary of aeronautics in each of them to underscore the continuing evolution of aeronautical research during the era of Apollo. I also wrote chapters 8 and 9, bringing developments in aeronautics and astronautics up to the present. In addition, many of the photos have been replaced, a short bibliographical essay was added, and the index has been revamped. As in the past, *Orders of Magnitude* was not intended as a definitive or interpretive study of the NACA and NASA. Even so, two recurrent themes can be discerned. One is the continuing relationship between NACA/NASA and the military services; another is the ongoing interaction with the European aerospace community.

I am grateful to many people who have cooperated in preparing the manuscript: Sylvia Fries, the NASA historian and Lee D. Saegesser, NASA archivist; and W. David Compton, a valued colleague, read and commented on the entire manuscript. Lee Saegesser also saved me from various errors of fact and turned up essential illustrations. At the History Office at NASA›s Johnson Space Center (JSC), I wish to thank Asha Vashi, Joey Pellarin, and Janet Kovacevich for supplying answers to many questions and for indefatigable good humor. Don Hess, who oversees the JSC History Office, facilitated access to JSC and its historical archives. At every NASA center, photo archivists and personnel in the Public Affairs offices provided necessary illustrations and information. Helen Heyder conscientiously typed different drafts of the entire manuscript. My family--Linda, Alex, and Paula--once again cheerfully endured the clutter of notes and books throughout our house.

In the process of defining the coverage and topics in this survey, I have been able to establish my own agenda, so that any shortcomings and errors are mine alone.

Roger E. Bilstein
Houston, Texas
1989

Chapter 1

NACA ORIGINS (1915-1930)

In 1915, Congressional legislation created an Advisory Committee for Aeronautics. The prefix "National" soon became customary, was officially adopted, and the familiar acronym NACA emerged as a widely recognized term among the aeronautics community in America.

The genesis of what came to be known as the National Advisory Committee for Aeronautics (NACA) occurred at a time of accelerating cultural and technological change. Only the year before, Robert Goddard began experiments in rocketry and the Panama Canal opened. Amidst the gathering whirlwind of the First World War, social change and technological transformation persisted. During 1915, the NACA's first year, Albert Einstein postulated his general theory of relativity and Margaret Sanger was jailed as the author of *Family Limitation*, the first popular book on birth control. Frederick Winslow Taylor, father of "Scientific Management," died, while disciples like Henry Ford were applying his ideas in the process of achieving prodigies of production. Ford produced his one millionth automobile the same year. In 1915, Alexander Graham Bell made the first transcontinental call, from New York to San Francisco, with his trusted colleague, Dr. Thomas A. Watson, on the other end of the line. Motion pictures began to reshape American entertainment habits, and New Orleans jazz began to make its indelible imprint on American music. At Sheepshead Bay, New York, a new speed record for automobiles was set, at 102.6 MPH, a figure that many fliers of the era would have been happy to match.

American flying not only lagged behind automotive progress, but also lagged behind European aviation. This was particularly galling to many aviation enthusiasts in the United States, the home of the Wright brothers. True, Orville and Wilbur Wright benefited from the work of European pioneers like Otto Lilienthal in Germany and Percy Pilcher in Great Britain. In America, the Wrights had corresponded with the well-known engineer and aviation enthusiast, Octave Chanute, and they had knowledge of the work of Samuel P. Langley, aviation pioneer and secretary of the Smithsonian Institution. But the Wrights made the first powered, controlled flight in an airplane on 17 December 1903, on a lonely stretch of beach near Kitty Hawk, North Carolina. Ironically, this feat was widely ignored or misinterpreted by the American press for many years, until 1908, when Orville made trial flights for the War Department and Wilbur's flights overseas enthralled Europe. Impressed by the Wrights, the Europeans nonetheless had already begun a rapid development of aviation, and their growing record of achievements underscored the lack of organized research in the United States.

Sentiment for some sort of center of aeronautical research had been building for several years. At the inaugural meeting of the American Aeronautical Society, in 1911, some of its members discussed a national laboratory with federal patronage. The Smithsonian Institution seemed a likely prospect, based on its prestige and the legacy of Samuel Pierpont Langley's dusty equipment, resting where it had been abandoned in his lab behind the Smithsonian "castle" on the Mall. But the American Aeronautical Society's dreams were frustrated by continued in-fighting among other organizations which were beginning to see aviation as a promising research frontier, including universities like the Massachusetts Institute of Technology, as well as government agencies like the U.S. Navy and the National Bureau of Standards.

Pre-World war I aviation technology. Military personnel struggle with a Wright biplane during trials at Fort Myer, Virginia, in 1908.

The difficulties of defining a research facility were compounded by the ambivalent attitude of the American public toward the airplane. While some saw it as a mechanical triumph with a significant future, others saw it as a mechanical fad, and a dangerous one at that. If anything, the antics of the "birdmen" and "aviatrixes" of the era tended to underscore the foolhardiness of aviation and airplanes. Fliers might set a record one month and fatally crash the next. Calbraith P. Rodgers managed to make the first flight from the Atlantic to the Pacific coast in 1911 (19 crashes, innumerable stops, and 49 days), but died in a crash just four months later. Harriet Quimby, the attractive and chic American aviatrix (she flew wearing a specially designed, plum-colored satin tunic), was the first woman to fly across the English Channel in 1912. Returning to America, she died in a crash off the Boston coast within three months.

There were fatalities in Europe as well, but the Europeans also took a different view of aviation as a technological phenomenon. Governments, as well as industrial firms, tended to be more supportive of what might be called "applied research." As early as 1909, the internationally known British physicist, Lord Rayleigh, was appointed head of the Advisory Committee for Aeronautics; in Germany, Ludwig Prandtl and others were beginning the sort of investigations that soon made the University of Gottingen a center of theoretical aerodynamics. Additional programs were soon under way in France and elsewhere on the continent. Similar progress in the United States was still slow in coming. Aware of European activity, Charles D. Walcott, secretary of the Smithsonian Institution, was able to find funds to dispatch two Americans on a fact-finding tour overseas. Dr. Albert F. Zahm taught physics and experimented in aeronautics at Catholic University in Washington, D.C.; Dr. Jerome C. Hunsaker, a graduate of the Massachusetts Institute of Technology, was developing a curriculum in aeronautical engineering at the institute. Their report, issued in 1914, emphasized the galling disparity between European progress and American inertia. The visit also established European contacts that later proved valuable to the NACA.

The outbreak of war in Europe in 1914 helped serve as a catalyst for the creation of an American agency. The use of German dirigibles for long-range bombing of British cities and the rapid evolution of airplanes for reconnaissance and for pursuit underscored the shortcomings of American aviation. Against this background, Charles D. Walcott pushed for legislative action to provide for aeronautical research allowing the United States to match progress overseas. Walcott received support from Progressive leaders in the country, who viewed government agencies for research as consistent with Progressive ideals such as scientific inquiry and technological progress. By the spring of 1915, the drive for an aeronautical research organization finally succeeded.

The enabling legislation for the NACA slipped through almost unnoticed as a rider attached to the Naval Appropriation Bill, on 3 March 1915. It was a traditional example of American political compromise.

As before, the move had been prompted by the Smithsonian. The legislation did not call for a national laboratory, since President Wilson apparently felt that such a move, taken during wartime conditions in Europe, might compromise America's formal commitment to strict nonintervention and neutrality. Although supported by the Smithsonian, the proposal emphasized a collective responsibility through a committee that would coordinate work already under way. The committee was an unpaid panel of 12 people, including two members from the War Department, two from the Navy Department, one each from the Smithsonian, the Weather Bureau, and the Bureau of Standards, and five more members acquainted with aeronautics. Despite concerns about appearing neutral, the proposal was tacked on as a rider to the naval appropriation bill as a ploy to clear the way for quick endorsement.

For fiscal 1915, the fledgling organization received a budget of $5000, an annual appropriation that remained constant for the next five years. This was not much even by standards of that time, but it must be remembered that this was an *advisory* committee only, "to supervise and direct the scientific study of the problems of flight, with a view to their practical solutions." Once the NACA isolated a problem, its study and solution was generally done by a government agency or university laboratory, often on an ad hoc basis within limited funding. The main committee of 12 members met semiannually in Washington; an Executive Committee of seven members, characteristically chosen from the main committee living in the Washington area, supervised the NACA's activities and kept track of aeronautical problems to be considered for action. It was a clubby arrangement, but it seemed to work.

In a wartime environment, the NACA was soon busy. It evaluated aeronautical queries from the Army and conducted experiments at the Navy yard; the Bureau of Standards ran engine tests; Stanford University ran propeller tests. But the NACA's role as mediator in the rancorous and complex dispute between Glenn Curtiss and the Wright-Martin Company represented its greatest wartime success. The controversy involved the technique for lateral control of aircraft in flight. Once settled, the resultant cross-licensing agreement consolidated patent rights and cleared the way for volume production of aircraft during the war as well as during the postwar era.

The authors of the NACA's charter had written it to leave open the possibility of an independent laboratory. Although several facilities for military research continued to function, the NACA pointed out in its first Annual Report for 1915 that civil aviation research would be in order when the Great War ended. And so, even before the war's conclusion, plans were afoot to acquire a laboratory. The best option seemed to be collaboration in the development of a new U.S. Army airfield, across the river from Norfolk, Virginia. The military facility was named after Samuel Pierpont Langley, former secretary of the Smithsonian; the NACA facility was named the Langley Memorial Aeronautical Laboratory, soon shortened to the familiar, cryptic "Langley."

Construction of the airfield got underway in 1917, hampered by the confusion following America's declaration of war on Germany and by the wet weather and marshy terrain of the Virginia tidewater region. One of the workers was an aspiring young writer named Thomas Wolfe. In his autobiograhical novel, Look Homeward Angel (1929), Wolfe's main character found a job at Langley as a horse-mounted construction supervisor paid $80 per month. He directed gangs striving to create a level airfield, pushing the earth "and filling interminably, ceaselessly, like the weary and fruitless labor of a nightmare, the marshy earth-craters, which drank their shovelled toil without end."

But eventually it did end; formal dedication took place on 11 June 1920. Although the Army, under wartime pressures, had already relocated its own research center to McCook Field, near Dayton, Ohio, Langley Field remained a large base, and military influence remained strong. The inaugural ceremonies included various aerial exhibitions and a fly-over of a large formation of planes led by the dashing Brigadier General William "Billy" Mitchell. Visitors found that the NACA's corner of Langley Field was comparatively modest: an atmospheric wind tunnel, a dynamometer lab, an administration building, and a small warehouse. There was a staff of 11 people--plenty of room to grow.

The Postwar Era

The management of the NACA and Langley, with a small staff for so many years, remained personal, straight-forward, and more or less informal. In Washington, a full-time executive secretary was named: John F. Victory, the NACA's first employee, hired in 1915. George W. Lewis, hired in 1919, became director of research, but remained in Washington, where he could palaver with politicians and joust with other bureaucrats. He spent long productive hours in the corridors of the Army-Navy Club and the Cosmos Club. Meanwhile, the close-knit staff down at Langley operated on a more democratic basis. In the lunchroom, junior staff, senior staff, and technicians dined together, where a free exchange of views continued over coffee and dessert. For years, Langley managed to attract the brightest young aeronautical engineers in the country, because they knew that their training would continue to expand by close and comradely contact with many senior NACA engineers on the cutting edge of research.

Engineers came to Langley from all over the country. Early employees often had degrees in civil or mechanical engineering, since so few universities offered a degree in aeronautical engineering alone. By the end of the 1920s, this had begun to change. From a handful of prewar courses dealing with aeronautical engineering, universities like the Massachusetts Institute of Technology evolved a plan of professional course work leading to both undergraduate and graduate degrees in the subject. The Daniel Guggenheim Fund for the Promotion of Aeronautics provided money for similar programs at several other schools. In 1929, a survey by an aviation magazine reported that 1400 aeroengineering students were enrolled in more than a dozen schools across the United States. The California Institute of Technology became a major beneficiary of the Guggenheim Fund's foresight. Although America possessed the facilities to train engineers and the NACA offered superb facilities for practical research, the country lacked a nerve center for advanced studies in theoretical aerodynamics. Germany led the world in this respect until the Guggenheim Fund lured the brilliant young scientist Theodore von Karman to the United States. Von Karman accepted a Caltech offer in 1929 and occupied his new post the following year. Within the decade, not only did Caltech's research projects enrich the field of aerodynamic theory, its graduates began to dominate the discipline in colleges and universities across the nation. The Guggenheim Fund's largesse was a tremendous stimulus to aeronautical engineering and research, as it was to the dozens of other aeronautical projects that it supported. Between 1926 and 1930, this personal philanthropy disbursed $3 million for a variety of fundamental research and experimental programs, including flight safety and instrument flying, that profoundly influenced the growth of American aviation.

Although the Langley organization became more formalized over time, there was maximum opportunity for individual initiative. The agency followed a regular procedure for instituting a "Research Authorization," but promising ideas could be pursued without formal approval. The NACA hierarchy in Washington and at Langley accepted this sort of "bootlegged" work as long as it was not too exotic, because it was often as productive as formal programs and kept the Langley staff moving out in front of the conventional frontier. The system also worked because the Langley staff remained small: about 100 in 1925. Creativity had its place, but outlandish projects were quickly spotted.

The sources for formal "Research Authorizations" were many and varied, often reflected by the catholic make-up of the NACA's main committee, drawing as it did from both military services, other government agencies, universities, and individuals from the aviation community. Ideas also came from Lewis's forays into Washington corridors of influence as well as from sources overseas. Edward Pearson Warner, serving as Langley's chief physicist, was packed off to Europe in 1920 to get a sense of postwar trends among major overseas countries; later the NACA set up a permanent observation post in Paris, where John J. Ide kept an eye on European activities up to World War II.

Langly Laboratory›s first wind tunnel, finished in 1920.

But research depended on facilities. At Langley, NACA technicians turned their attention to a new wind tunnel. It was not large, designed to have a test section of about five feet in diameter, but it could be configured to produce speeds of 120 MPH in the test section, making it one of the best facilities in the world. Still, there were inherent drawbacks. With no firsthand experience, NACA planners built a conventional, open circuit tunnel based on a design proven at the British National Physical Laboratory. At the University of Gottingen in Germany the famous physicist Ludwig Prandtl and his staff had already built a closed circuit, return-flow tunnel in 1908. Among other things, the closed circuit design required less power, boasted a more uniform airflow, and permitted pressurization as well as humidity control.

The NACA engineers at Langley knew how to scale up data from the small models tested in their sea level, open circuit tunnels, but they soon realized that their estimates were often wide of the mark. For significant research, the NACA experimenters needed facilities like the tunnels in Gottingen. They also needed someone with experience in the design and operation of these more exotic tunnels. Both requirements were met in the person of Max Munk.

Munk had been one of Prandtl's brightest lights at Gottingen. During World War I, many of Munk's experiments in Germany were instantaneously tagged as military secrets (though they usually appeared in England, completely translated, within days of his completing them). After the war, Prandtl contacted his prewar acquaintance, Jerome Hunsaker, with the news that Munk wanted to settle in America. For Munk to enter the United States in 1920, President Woodrow Wilson had to sign two special orders: one to get him into America so soon after the war, and one permitting him to hold a government job. In the spring of 1921, construction of a pressurized, or variable density tunnel, began at Langley. The goal was to keep using models in the tunnel, but conduct the tests in a sealed, airtight chamber where the air would be compressed "to the same extent as the model being tested." In other words, if a one-twentieth scale model was being tested in the variable density tunnel, then researchers would increase the density of air in the tunnels to a level of 20 atmospheres. Results could be expressed in a numerical scale known as the Reynolds number. The tunnel began operations in 1922 and proved highly successful in the theory of airfoils. As one Langley historian wrote, "Langley's VDT (variable density tunnel) had established itself as the primary source for aerodynamic data at high Reynolds numbers in the United States, if not in the world." Munk's tenure at the NACA was a stormy one. He was brilliant, erratic, and an autocrat. After many confrontations with various bureaucrats and Langley engineers, Munk resigned from the NACA in 1929. But his style of imaginative research and sophisticated wind tunnel experimentation was a significant legacy to the young agency.

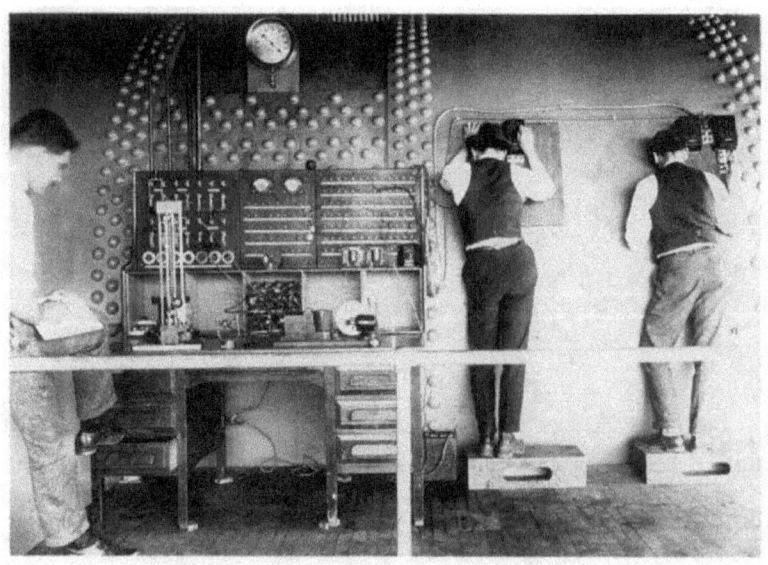

A NACA team conducts research using the variable density tunnel in 1929.

The variable density tunnel, using scale models, represented only one avenue of aeronautical investigation. In parallel, the NACA ran a program of full scale flight tests that also yielded early dividends. In the process, the NACA helped establish a body of requisite guidelines and procedures for flight testing. One problem involved instrumentation--proper equipment for acquiring accurate data on full scale aircraft during actual flight that could correlate with data obtained in wind tunnels. In one early project, wind tunnel data for a model of the Curtiss JN-4 "Jenny" was compared to information derived from an instrumented Jenny put through a series of flight tests to investigate lift and drag. By comparing data, the reliability of wind tunnel information could be judged more rigorously. The tests of the 100 MPH JN-4 represented the start of carefully planned and instrumented experimental flights that became a hallmark of the NACA and NASA from subsonic through supersonic flight. The early JN-4 flights also uncovered another aspect of flight testing to be addressed--the need for specially trained test pilots. Langley also pioneered in the concept of training fliers as test pilot-engineers.

By 1922, several different kinds of aircraft were under test at Langley. Three workhorse planes were Curtiss JN-4H Jennies, used for a series of takeoff and landing and performance measurements that represented an important new set of design parameters. Military investigations also began during these early years, when the Navy Bureau of Aeronautics came to the NACA for a comparative study of airplanes in terms of stability, controllability, and maneuverability. Along with a Vought VE-7 from the Navy, Langley pilots obtained a Thomas-Morse MB-3 from the Army, and two foreign models: a British SE-5A (one of the Royal Air Force's principal fighters of World War I) and a German Fokker D-VII (the main source of references to the "Fokker scourge" during the war). Evaluating front-line aircraft from foreign as well as American air forces inaugurated a practice that persisted through the NASA era as well. Other investigations during the mid-1920s involved further work for the Navy, to ascertain accurate data on stall, takeoff, and landing speeds of a specific aircraft. The Army turned up with a similar request for studies of these and other qualities for most of the aircraft in the Air Service inventory at that time.

The progressive experience in flight test work, including a variety of instrumentation required to register the data, contributed to studies of pressure distribution along wing surfaces, a major effort during the 1920s. Beginning with measurements during steady flight, test pilots and instrumentation experts devised techniques to study pressure distribution during accelerated flight and in maneuvers, accumulating invaluable design data where none had existed before. Steady improvement in instrumentation permitted pressure distribution surveys to be wound up in one day, rather than making a prolonged series of flights lasting as long as two months. By 1925, Langley had 19 aircraft dedicated to a variety of test operations. Ground testing had expanded to include a new engine research laboratory in which engineers had begun work on supercharging of engines for high altitude bombers, as well as a means of boosting power for interceptors in order to give them a high rate of climb--the sort of investigative work that paid dividends later in World War II.

The Tunnels Pay Off

In the meantime, the variable density tunnel began to pay further dividends in the form of airfoil research. During the late 1920s and into the 1930s, the NACA developed a series of thoroughly tested airfoils and devised a numerical designation for each airfoil--a four digit number that represented the airfoil section's critical geometric properties. By 1929, Langley had developed this system to the point where the numbering system was complemented by an airfoil cross-section, and the complete catalog of 78 airfoils appeared in the NACA's annual report for 1933. Engineers could quickly see the peculiarities of each airfoil shape, and the numerical designator ("NACA 2415," for instance) specified camber lines, maximum thickness, and special nose features. These figures and shapes transmitted the sort of information to engineers that allowed them to select specific airfoils for desired performance characteristics of specific aircraft.

During the late 1920s, the NACA also announced a major innovation that resulted in the agency's first Robert J. Collier Trophy, presented annually by the National Aeronautic Association for the year's most outstanding contribution to American aviation. In 1929, the Collier trophy went to the NACA for the design of a low-drag cowling.

Most American planes of the postwar decade mounted air-cooled radial engines, with the cylinders exposed to the air stream to maximize cooling. But the exposed cylinders also caused high drag. Because of this, the U.S. Army had adopted several aircraft with liquid-cooled engines, in which the cylinders were arranged in a line parallel to the crankshaft. This reduced the frontal area of the aircraft and also allowed an aerodynamically contoured covering, or nacelle, over the nose of the plane. But the liquid-cooled designs carried weight penalties in terms of the myriad cooling chambers around the cylinders, gallons of coolant, pumps, and radiator. The U.S. Navy decided not to use such a design because the added maintenance requirements cut into the limited space aboard aircraft carriers. Moreover, the jarring contact of airplanes with carrier decks created all sorts of cracked joints and leaks in liquid-cooled engines. Air-cooled radial engines simplified this issue, although their inherent drag meant reduced performance. In 1926, the Navy's Bureau of Aeronautics approached the NACA to see if a circular cowling could be devised in such a way as to reduce the drag of exposed cylinders without creating too much of a cooling problem.

While significant work on cowled radial engines proceeded elsewhere, particularly in Great Britain, investigations at Langley soon provided a breakthrough. American aerodynamicists at this time had the advantage of a new propeller research tunnel completed at Langley in 1927. With a diameter of 20 feet, it was possible to run tests on a full-sized airplane. Following hundreds of tests, a NACA technical note by Fred E. Weick in November 1928 announced convincing results. At the same time, Langley acquired a Curtiss Hawk AT-5A biplane fighter from the Air Service and fitted a cowling around its blunt radial engine. The results were exhilarating. With little additional weight, the Hawk's speed jumped from 118 to 137 MPH, an increase of 16 percent. The virtues of the NACA cowling received

A Sperry Messenger mounted for testing in Langley's propeller research tunnel in 1927.

public acclaim the next year, when Frank Hawks, a highly publicized stunt flier and air racer, added the NACA cowling to a Lockheed Air Express monoplane and racked up a new Los Angeles/New York nonstop record of 18 hours and 13 minutes. The cowling had raised the plane's speed from 157 to 177 MPH. After the flight, Lockheed Aircraft sent a telegram to the NACA committee: "Record impossible without new cowling. All credit due NACA for painstaking and accurate research." By using the cowling, the NACA estimated savings to the industry of over $5 million--more than all the money appropriated for NACA from its inception through 1928.

The NACA cowling, as fitted on a Curtiss Hawk, a standard U.S. Army combat plane.

After 15 years, the sophistication of the NACA's research had dramatically changed. And so had the sophistication of aviation. After a fitful start in 1918, the U.S. government's airmail service had forged day-and-night transcontinental routes across America by 1924. The service saved as much as two days in delivering coast-to-coast mail, accelerating the tempo of a business civilization and saving millions of dollars. In 1925, the government began to contract for service with privately owned companies, a change that marked the beginning of the airline industry. By the end of the decade, the private companies were beginning to fly passengers as well as mail, and Pan American Airways had launched international services between Florida and Cuba, as well as between Texas and Central America. Following the Air Commerce Act of 1926, lighted airways were improved, radio communications progressed, and guidelines were established for pilot proficiency as well as aircraft design and construction. By the time Charles Lindbergh made his solo flight from New York to Paris in 1927, an aeronautical infrastructure was already in place. The "Lindbergh Boom" that followed his striking achievement could not have been sustained without the important progress of the previous years.

The NACA helped spur much of this development through its refinement of wing design and investigations of various aerodynamic phenomena. The agency also benefited from overall aviation progress during this era, sharing the increased aviation budgets represented by funds for civil programs under the Air Commerce Act and for the expansion of U.S. Army and U.S. Navy aviation. The Army Air Service was granted more autonomy in 1926, when it became the Air Corps. During the 1920s, the Army's air arm began to develop a doctrine, standardize its training, and pursue advanced research, often in cooperation with the NACA. In the development of equipment, the Air Service undertook projects for modern fighters and strategic bombers to come. The U.S. Navy experienced similar organizational changes and began the construction and operational evaluation of aircraft carriers, like the Langley, Lexington, and Saratoga.

Collectively, the progress of civilian aviation, military aviation, and aeronautical research set the stage for the aeronautical revolution that began in the 1930s. The design characteristics of the 1920s--fabric covered biplanes with radial engines--gave way to truly sophisticated airplanes of the 1930s with streamlined shapes, metal construction, retractable landing gear, and high performance. The national economy may have sagged during the Great Depression of the 1930s, but the aviation industry reached new levels of excellence.

Early Rocketry

There were some areas of flight technology, such as rocketry, in which the NACA did not become involved. Nevertheless, when the NACA was transformed into NASA in 1958, the new space agency could reach back into some forty years of American and European writing and research on rocketry and the possibilities of space flight. During the 1920s, the subject of space flight more often seemed to be the province of cranks and science fiction writers spinning wildly improbable tales. But visionary researchers in the United States, as well as Great Britain, Germany, Russia, and elsewhere were taking the first hesitant steps toward actual space travel. In America, Robert Hutchings Goddard is remembered as one of the foremost pioneers.

After completing a doctorate in physics at Clark University in 1911, Goddard joined its faculty. During his physics lectures, he sometimes startled students by outlining various ways of reaching the Moon. Despite the students' skepticism, Goddard was basing his projections on the very real advances in metallurgy, thermodynamics, navigational theory, and control techniques. Twentieth century technology had begun to make rocketry and space flight feasible. Goddard fabricated a series of test rockets, and in 1920 wrote a classic monograph, A Method of Attaining Extreme Altitudes, published by the Smithsonian. In it, he described how a small rocket could soar from the Earth to the Moon, and detonate a payload of flash powder on impact, so that observers using large telescopes on Earth could verify the rocket's arrival on the lunar surface. Caustic news stories about rocketry and lunacy caused Goddard, a shy individual, to shun publicity during the remainder of his life.

Goddard continued to experiment with liquid propellant rockets, igniting them in a field on his Aunt Effie's farm, where their piercing screeches disturbed the neighbor's livestock. Eventually, on 16 March 1926, one of Goddard's devices lifted off to make the first successful flight of a liquid propellant rocket.

Robert H. Goddard, with the first successful liquid-fuel chemical rocket, launched 16 March 1926.

At the time, it was hardly an earthshaking demonstration--a flight of 2.5 seconds that carried the rocket to an altitude of 41 feet. A small, but significant step towards future progress. Continued work caught the attention of Charles Lindbergh, who persuaded the Guggenheim Fund to support Goddard's research. By the 1930s, Goddard set up shop at a desert site near Roswell, New Mexico, where he and a small group of assistants developed liquid propellant rockets of increasing size and complexity. Unfortunately, Goddard's reticence meant that he labored in isolation, and other experimental groups knew little of his activities. "His own penchant for secrecy set him apart from the mainstream," wrote historian Frank Winter. "As a result, Goddard's monumental advances in liquidfuel technology were largely unknown until as late as 1936 when his second Smithsonian report, Liquid Propellant Rocket Development appeared." In the meantime, researchers in Germany began work that eventually had an impact on the American space program.

Rocket enthusiasts in Germany took inspiration from the same science fiction (Jules Verne and others) that had motivated Goddard and took advantage of advances in metallurgy and chemistry. They also took another important step, establishing an organization that facilitated the exchange of information and accelerated the rate of experimentation. In 1927, the Verein fur Raumschiffart (VfR) was founded by Hermann Oberth and others. A year later, the VfR collaborated with producers of a science fiction film on space travel, The Girl in the Moon. The script included the now-famous countdown sequence before ignition and lift-off. For publicity, the VfR hoped to build and launch a small rocket. The rocket project fizzled, but among the design team was an eager 18-year-old student named Wernher von Braun, whose enthusiasm for space flight never waned.

In Russia, Konstantin Tsiolkovsky left a legacy of significant writing in the field of rocketry. Although Tsiolkovsky did not construct any working rockets, his numerous essays and books helped point the way to practical and successful space travel. Tsiolkovsky spent most of his life as an unknown mathematics teacher in the Russian provinces, where he made some pioneering studies in liquid chemical rocket concepts and recommended liquid oxygen and liquid hydrogen as the optimum propellants. In the 1920s, Tsiolkovsky analyzed and mathematically formulated the technique of staging vehicles to reach escape velocities from Earth. Rocket societies were organized as early as 1924 in the Soviet Union, but the barriers of distance and politics limited interchange between these groups and their western counterparts. In 1931, the Group for the Study of Reaction Motion, known by its Russian acronym of GIRD, became organized, with primary research centers in Moscow and Leningrad. The activity by GIRD resulted in the Soviet Union's first liquid-fuel rocket launch in 1933. Although GIRD stimulated considerable activity in the Soviet Union, including conferences, periodicals, and hardware development, military influences became increasingly dominant. The devastating purges of the 1930s seem to have decimated the astronautical leadership in the Soviet Union, so that the rapid recovery of Soviet activity in the postwar era was all the more remarkable.

In many ways, astronautics became professionalized, much as aeronautics. The term "astronautics" also became more commonplace. The designation grew out of a dinner meeting in Paris in 1927. A Belgian science fiction author, J. J. Rosny, came up with the word, which was then popularized by the French writer and experimenter, Robert Esnault-Pelterie, whose best-known book, L'Astronautique, appeared in 1930. With a body of literature, evolving technology, active professionals, and an identity, astronautics--like aeronautics--was poised for rapid growth.

Chapter 2

NEW FACILITIES, NEW DESIGNS (1930-1945)

To many NACA engineers, the agency's first fifteen years represented remarkable aeronautical progress. The next fifteen years, from 1930 to 1945, seemed even more remarkable, as streamlined aircraft became commonplace, World War II spawned an impressive variety of modern combat planes, and rocketry became an awesome force in twentieth century warfare.

The propeller research tunnel at Langley continued to yield significant information that resulted in equally significant design refinements in the new generation of airplanes. One of the most obvious had to do with fixed landing gear. As a means to increase speed, retractable landing gear was not unknown, since this approach had been tried on various airplanes before and after World War I. But retractable gear required additional equipment for raising and lowering and appeared to lack the ruggedness and reliability of conventional, fixed gear. On the other hand, fixed gear was thought to be a major drag factor, although nobody had accurately assessed the aerodynamic liability. NACA engineers set up a series of tests using the propeller research tunnel to get an accurate measure of the fixed gear's drag on a Sperry Messenger. The results were astonishing. Fixed gear was estimated to create nearly 40 percent of the total drag acting on the plane. This eye-opening news, a dramatic demonstration of the performance penalty incurred by fixed gear, prompted rapid development of retractable gear for a wide variety of airplanes. The NACA's tests played a large role in the evolution of modern, retractable-geared aircraft.

There were further projects that pointed the way to sleeker airplanes emerging by the end of the 1930s. Tri-motored airliners, like the Fokkers, Fords, and Boeings, had become standard equipment in America and elsewhere during the late 1920s. They could not easily be redesigned to mount retractable gear, but the trio of big, blunt radial engines that powered them could be shrouded with the new NACA cowling to give them much improved performance. Engineers at Langley took a Fokker trimotor powered by three Wright J-5 Whirlwind engines and fitted it with cowlings. Confident expectations of sudden enhancement of performance were dashed and engineers were baffled. They began to wonder if the installation of engines had something to do with it. So as not to encumber the wing, the original designers had placed the engines on struts beneath the wing (or, in the case of biplanes like the Boeing 80, between the wings). After getting the big Fokker set up in the propeller research tunnel, Langley engineers ran a series of tests that conclusively changed the looks of multiengine transports to come. They discovered that the best position for the engines was neither above or below the wing, but mounted as part of its structure--situated ahead of the wing, with the engine nacelle faired into the wing's leading edge.

This was the sort of information that also contributed to the evolution of the modern airliners of the decade. Conventional wisdom in the past had dictated that wings should be mounted high on the fuselage, permitting engines to be slung underneath with clearance for the propeller arc. This meant complex struts (creating drag) and led to the use of awkward, long-legged fixed gear (creating even more drag). By mounting engines in the wing's leading edge, the wing could be positioned on the lower part of the fuselage, which meant that the landing gear was now short-legged and less awkward--in fact, retractable. Influenced by NACA research, low- winged monoplanes with retractable gear soon replaced the high-winged design for airliners and many other aircraft.

The propeller research tunnel at Langley had obviously been a profitable facility, although it had limitations for thorough testing of full-sized aircraft. In 1931, when the full scale tunnel was officially dedicated, Langley engineers used it to launch a new round of evaluations which, while sometimes less dramatic than cowlings, unquestionably added new dimensions to the science of aerodynamics. Its impressive statistics marked the beginning of test facilities of heroic proportions.

Nonetheless, the full scale tunnel did not overshadow other Langley test facilities. There were those who felt that the shortcomings of the variable density tunnel, with its acknowledged drawbacks in turbulence, would soon be eclipsed by the huge full scale tunnel. With partisans on both sides, friction between personnel from the variable density tunnel and the full scale tunnel became legendary. In time, both established a relevant niche in the scheme of things. Meanwhile, the variable density tunnel played a key role in many projects, and its personnel made a singular contribution to the theory of the laminar flow wing.

While the variable density tunnel could test many more varieties of aircraft designs, which could be built as scale models, the turbulence issue continued to dog research findings. In the process of studying this issue, researchers took a closer look at flow phenomena, especially the "boundary layer," where so many problems seemed to crop up. The boundary layer was known to be a thin structure of air only a few thousandths of an inch from the contour of the airfoil. Within it, air particles changed from a smooth laminar flow from the leading edge to a more turbulent state towards the trailing edge. In the process, drag increased. After observing tests in a smoke tunnel and evaluating other data, aerodynamicists concluded that the prime culprits in disrupting laminar flow were traceable to the wing's surface (rivet heads and other rough areas) and to pressure distribution over the wing's surface.

A Vought O3U set up for tests using the full scale wind tunnel at Langley, completed in 1931.

Eastman Jacobs, head of the variable density tunnel section, came up with various formulas to allow for the tunnel's turbulence in evaluating models and pushed for a larger, improved tunnel. He also championed a systematic experimental approach in airfoil development.

Jacobs was often challenged by a Norwegian emigre, Theodor Theodorsen, of the Physical Research Division. Theodorsen, steeped in mathematical research, was a strong proponent of airfoil investigation by theoretical study. His opposition to Jacobs's proposal for an improved variable density tunnel and his insistence that, instead, Langley personnel needed more mathematical skills and theoretical concepts, sharpened the debate between experimentalists and theorists within the NACA. Jacobs, in fact, kept abreast of current theories, and he eventually fashioned a theoretical approach, backed up by his trademark experimental style that led to advanced laminar flow airfoils.

While the NACA deserves credit for its eventual breakthrough in laminar flow wings, the resolution of the issue illustrates a fascinating degree of universality in aeronautical research. The NACA--born in response to European progress in aeronautics-- benefited through the employment of Europeans like Munk and Theodorsen, and profited from a continuous interaction with the European community.

In 1935, Jacobs went to Rome as the NACA representative to the Fifth Volta Congress on High-Speed Aeronautics. During the trip, he visited several European research facilities, comparing equipment and discussing the newest theoretical concepts. The United States, he concluded, held a leading position, but he asserted that "we certainly cannot keep it long if we rest on our laurels." On his way home, Jacobs stopped off at Cambridge University in Great Britain for long visits with colleagues who were investigating the peculiarities of high-speed flow, including statistical theories of turbulence. These informal exchanges proved to be highly influential on Jacobs' approach to the theory of laminar flow by focusing on the issue of pressure distribution over the airfoil. Working out the details of the idea took three years and engaged the energies of many individuals, including several on Theodorsen's staff, even though he remained skeptical.

Once the theory appeared sound, Jacobs had a wind tunnel model of the wing rushed through the Langley shop and tested it in a new icing tunnel that could be used for some low-turbulence testing. The new airfoil showed a fifty percent decrease in drag. Jacobs was elated, not only because the project incorporated complex theoretical analysis, but also because the subsequent empirical tests justified a new variable density tunnel.

In application, the laminar flow airfoil was used during World War II in the design of the wings for the North American P-51 Mustang, as well as some other aircraft. Operationally, the wing did not enhance performance as dramatically as tunnel tests suggested. For the best performance, manufacturing tolerances had to be perfect and maintenance of wing surfaces needed to be thorough. The rush of mass production during the war and the tasks of meticulous maintenance in combat zones never met the standards of NACA laboratories. Still, the work on the laminar flow wing pointed the way to a new family of successful high-speed airfoils. These and other NACA wing sections became the patterns for aircraft around the world.

The NACA's laminar flow airfoil was first used on the North American XP-51 Mustang.

NACA reports began to emerge from an impressive variety of tunnels that went into operation during the 1930s. The refrigerated wind tunnel, declared operational in 1928, became a major tool for the study of ice formation on wings and propellers. In flight, icing represented a menace to be prevented at all costs. Langley's research in the refrigerated tunnel contributed to successful deicing equipment that not only enabled airliners to keep better schedules in the 1930s but also enabled World War II combat planes to survive many encounters with bad weather. Another facility at Langley, a free-spin wind tunnel, yielded vital information on the spin characteristics of many aircraft, improving their maneuverability while avoiding deadly spin tendencies. A hydrodynamics test tank solved many riddles for designers of seaplanes and amphibians, by towing hull models to simulated takeoff speeds.

The NACA also took a bold look ahead to much higher airplane speeds to come. In the mid-1930s, when speeds of 200 MPH were quite respectable, the agency proposed a "fullspeed" tunnel, providing the means for tests at a simulated 500 MPH. With an 8 foot diameter, the tunnel allowed tests of comparatively large models, as well as some full scale components. Completed early in 1936, the eight- foot tunnel played a major role in high-speed aerodynamic research, laying the foundations for later work in high subsonic speeds as well as the baffling transonic region.

As the research capabilities of the NACA expanded, so did the persistent, nagging problems that followed the introduction of successive generations of aircraft. For the NACA, one of the most unusual apparitions to appear in the 1930s was the autogyro. First developed by a Spaniard, Juan de la Cierva, in the 1920s, the autogyro was thought to have great promise in the immediate future. At first glance, it looked like a helicopter, with a huge multi- bladed rotor situated above the fuselage. Unlike the helicopter, the autogyro had stubby wings and used a nose-mounted engine with a conventional propeller for forward momentum. In moving ahead, the main rotor turned, so that its long thin airfoil blades provided lift, with some assistance from the shortened wings. The autogyro could not take off or land vertically, nor could it hover, but its abbreviated landing and takeoff runs were dramatic, and proponents claimed that the aircraft minimized dangerous stalls. Some writers of the era envisioned the autogyro as a replacement for the family sedan. Accordingly, the NACA bought a Pitcairn PCA-2 autogyro (designed and manufactured in Pennsylvania by Harold Pitcairn) and began tests in 1931. These trials did not contribute to a permanent niche in American life for the autogyro, but Langley was launched into continuing work on rotary-wing aircraft. In fact, some of the maneuverability tests and other investigations on the autogyro led to testing criteria used into the 1980s.

Flight research like that involving the autogyro marked this activity as an increasingly valued component of Langley's procedures. Accomplished on an ad hoc basis most of the time, flight testing became more formalized in 1932, when a flight test laboratory appeared at Langley. With separate space allocated for staff, shop work, and an aircraft hangar, the new laboratory made its own contributions to aviation progress during the 1930s.

Expanded flight test operations included evaluation of the Pitcairn autogyro.

Among the various airplanes that passed through Langley were two of the most advanced airliners of the era: the Boeing 247 and the Douglas DC-1, which led to the classic DC-3. The Boeing and Douglas designs incorporated the latest aviation technology that had evolved since the end of World War I. With the Ford Tri-Motor of the 1920s, wooden frame and fabric covering had given way to all-metal construction. Unlike the Ford, the Boeing and Douglas transports were low-winged planes with retractable landing gear, and their more powerful twin engines were cowled and mounted into the leading edge of the wings. At 170-180 MPH, they were considerably faster than any of their counterparts, and attention to details like soundproofing and other passenger comforts made them far more popular with travelers. Later versions of the Douglas transport, like the DC-3, added refinements like wing flaps and variable pitch propellers that made it even more effective in takeoffs and landings, as well as cruising at optimum efficiency at higher altitudes. But it was not clear what would happen if one of the two engines on the new transports failed. At the request of Douglas Aircraft, Langley evaluated problems of handling and control of a twin-engine transport with one engine out. These tests, conducted just six months before the DC-3 made its maiden flight, provided the sort of procedures to allow pilots to stay aloft until an emergency landing could be made.

The design revolution leading to all-metal monoplane transports had a similar impact on military aircraft. During 1935, Boeing began flight tests of its huge, four-engined Model 299, the prototype for the B-17 Flying Fortress of the Second World War. The big airplane's performance exceeded expectations, due in no small part to design features pioneered by the NACA. The Boeing Company sent a letter of appreciation to the NACA for specific contributions to design of the plane's flaps, airfoil, and engine cowlings. The letter concluded, "it appears your organization can claim a considerable share in the success of this particular design. And we hope that you will continue to send us your 'hot dope' from time to time. We lean rather heavily on the Committee for help in improving our work."

The ability of the NACA to carry out the sort of investigations that proved useful was often the result of continuing contacts with the aviation community. One of the most interesting formats for such ideas was the annual aircraft engineering conference, which began in 1926. Attendees included the movers and shakers from the armed services, the aviation press, government agencies, airlines, and manufacturers. These were busy people, and the NACA gave them a carefully orchestrated two-day visit to Langley, with plenty of time for conversation.

Over 300 people made each annual trip, an invitation only opportunity during the 1930s. The NACA's executive secretary, John Victory, became the principal organizer of the event, which had almost sybaritic overtones in a depression era. After gathering in Washington, the group boarded a chartered steamer for a stately cruise down the Chesapeake Bay to Hampton, Virginia. Once ashore, the travelers partook of a generous Southern breakfast at a local resort hotel, then headed for Langley in an impressive motorcade that numbered over 50 cars. The program included reviews of current projects, followed by smaller group tours, lab demonstrations, and technical sessions throughout the day. Conference participants motored back to the hotel for cocktails on the veranda, an elaborate banquet, and an overnight return cruise to Washington. Public relations played an obvious role in such outings, but the conferences represented a useful avenue for maintaining contact, for keeping a finger on the pulse of the aviation community, and for keeping the aviation community abreast of the NACA's latest research and facilities.

Although the NACA personnel may not have enjoyed luxurious perquisites on a daily basis, the agency continued to be a magnet for many young aeronautical engineers. Langley's impressive facilities in particular were a powerful lure, in addition to the opportunity to work closely with well-known people at the cutting edge of flight. Through the 1930s, Langley managed to maintain a degree of informality that provided a unique environment for newly hired personnel. John Becker, who reported for duty in 1936, remembered the crowded lunchroom where he found himself rubbing shoulders with the authors of NACA papers he had just been studying at college. "These daily lunchroom contacts provided not only an intimate view of a fascinating variety of live career models," he wrote, "but also an unsurpassed source of stimulation, advice, ideas, and amusement." The tables in the lunchroom had white marble tops. By the end of the lunch hour, the table tops were invariably covered by sketches, equations, and other miscellany, erased by hand or by a napkin and drawn over again. Becker lamented the loss of this "great unintentional aid to communication" when Langley's growing staff required a larger, modern cafeteria with unusable table surfaces.

Much of this growth--and the end of an era for Langley and the NACA--occurred during the wartime period. In 1938, the total Langley staff came to 426. Just seven years later, in 1945, Langley numbered 3000 personnel.

Military Research

The prewar research at Langley had a catholic fallout, in that the center's activities were applicable to both civil and military aircraft. The commercial aircraft and fighting planes of the first one-and-a-half decades following World War I were very similar in terms of airspeed, wing loading, and general performance. For example, Langley's work on the cowling for radial engines had the encouragement of both civil and military personnel, and the NACA cowling eventually appeared on a remarkable variety of light planes, airliners, bombers, and fighter aircraft. Many other NACA projects on icing, propellers, and so on were equally useful to civil and military designs.

About the mid-1930s the phenomenon of mutual benefits began to change. Commercial airline operators put a premium on safety and operational efficiency. While such factors were not shunned by military designers, the qualities of speed, maneuverability, and operations to very high altitudes meant that NACA research increasingly proceeded along two separate paths. By 1939, the Annual Manufacturers Conference was phased out and replaced by an "inspection," planned solely for representatives of the armed services and delegates from firms having military contracts.

For most of the time after the mid-1930s benchmark, military R&D took the lead in the NACA, and its fallout was incorporated into civilian airplanes. Moreover, there are indications that the U.S. Navy often fared better than the U.S. Army in reaping benefits from Langley's extensive R&D talents. This situation may have stemmed from Langley's early days, when there was some friction about civilian NACA facilities located at the Army's Langley Field. Old hands at the NACA felt that certain Army people wanted to shift the NACA's work to McCook Field in Ohio and to conduct all of its operations under an Army umbrella. Under the circumstances, the Navy appeared to have smoother relations with the NACA. At the same time, the Navy had reason to rely heavily on the NACA's expertise. During the 1920s and 1930s, the service developed its first aircraft carriers. Concurrently, a rather special breed of aircraft had to be developed to fit the demanding requirements of carrier operations. Landings on carriers were bone-jarring events repeated many times (a carrier landing was wryly described as a "controlled crash"); takeoffs were confined to the limited length of a carrier's flight deck. In the process of beefing up structures, improving wing lift, keeping aircraft weight down, enhancing stability and control, and studying other problems, naval aviation and the NACA grew up together. Between 1920 and 1935, the Navy submitted twice as many research requests as the Army.

There were still some instances in which civilian needs benefited military programs. In 1935, Edward P. Warner, Langley's original chief physicist, was working as a consultant for the Douglas Aircraft Company. Warner had the job of determining stability and control characteristics of the DC-4 four-engined transport. Accepted practice of the day usually meant informal discussions between pilots and engineers as the latter tried to design a plane having the often elusive virtues of "good flying qualities." At Warner's request the NACA began a special project to investigate flying qualities desired by pilots so that numeric guidelines could be written into design specifications. At Langley, researchers used a specially instrumented Stinson Reliant to develop usable criteria. Measurable control inputs from the test pilot were correlated with the plane's design characteristics to develop a numeric formula that could be applied to other aircraft. Further tests on 12 different planes gave a comprehensive set of figures for both large and small aircraft. As military programs gained urgency in the late 1930s, the formulas for flying qualities were increasingly used in the design of new combat planes.

The growing international threat found the American aviation industry in far better shape than was the case on the eve of World War II. In terms of civil aviation, the United States had established an enviable record of progress. Commercial airliners like the DC-3 had set a world standard and, in fact, were widely used by many foreign airlines on international routes. Airline operations had reached new levels of maturity, not only in terms of marketing and advertising to attract a growing clientele, but also in a myriad variety of supporting activities. These included maintenance and overhaul procedures, radio communication, weather forecasting, and long-distance flying. Many of these skills proved valuable to the military after the outbreak of war. Pan American World Airways (Pan Am), which had pioneered long distance American routes throughout the Caribbean, Pacific, and Atlantic shared its skills and personnel to help the Air Transport Command evolve a remarkable global network during the war years. Pan Am relied on a series of impressive flying boats designed and built by Sikorsky, Martin, and Boeing during the 1930s. Although the military airlift services depended more on land-planes like the DC-3 (military version known as the C-47) and DC-4 (or C-54), many of the imaginative design concepts of the flying boats pointed the way to the multi-engined airliners that replaced them.

There were even benefits for the light plane industry. Despite the depression, personal and business flying became firmly entrenched in the American aviation scene. Manufacturers offered a surprising array of designs, from the economical two-place Piper Cub J-3 to the swift 45 place business planes produced by Stinson and Cessna. At the top of the scale the Beech D-18, a twin-engine speedster, offered the era's ultimate in corporate transportation. When war came, these and other manufacturers were ready to turn out the dozens of primary trainers (larger planes for navigational and bombing instruction) and various components that made up the other equipment in the U.S. armed forces.

The Air Force itself was beginning to receive the sort of combat planes that enabled it to meet aggressive fliers in the skies over Europe and the Far East. Prewar fighters like the Curtiss P-40 soon gave way to the Lockheed P-38, Republic P-47, and North American P-51. A new family of medium bombers and heavy bombers included the redoubtable B-17 Flying Fortress, derived from the Boeing 299. Aboard the U.S. Navy's big new aircraft carriers, biplanes had given way to powerful monoplanes like the Grumman Wildcat, followed by the Hellcat and Vought Corsair. There were also new dive bombers and long-legged patrol planes like the Catalina amphibian. Directly or indirectly, the majority of these aircraft profited from the NACA's productivity during the 1930s as well as during the war.

The War Years

Even though Langley and the NACA had contributed heavily to the progress of American aviation, there were still some in Congress who had never heard of them. Before World War II, a series of committee reports brought a dramatic change. During the late 1930s, John Jay Ide, who manned NACA's listening post in Europe, reported unusually strong commitments to aeronautical research in Italy and Germany, where no less than five research centers were under development. Germany's largest, located near Berlin, had a reported 2000 personnel at work, compared to Langley's 350 people. Although the Fascist powers were developing civil aircraft, it became apparent that military research absorbed the lion's share of work at the new centers. Under the circumstances, the NACA formed stronger alliances with military services in the United States for expansion of its own facilities.

In 1936, the agency put together a special committee on the relationship of NACA to National Defense in time of war, chaired by the Chief of the Army Air Corps, Major General Oscar Westover. Its report, released two years later, called for expanded facilities in the form of a new laboratory--an action underscored by Charles Lindbergh, who had just returned from an European tour warning that Germany clearly surpassed America in military aviation. A follow-up committee, chaired by Rear Admiral Arthur Cook, chief of the Navy's Bureau of Aeronautics, recommended that the new facility should be located on the West Coast, where it could work closely with the growing aircraft industry in California and Washington. Following congressional debate, the NACA received money for expanded facilities at Langley (pacifying the Virginia Congressman who ran the House Appropriations Committee) along with a new laboratory at Moffett Field, south of San Francisco. The official authorization came in August 1939; only a few weeks later, German planes, tanks, and troops invaded Poland. World War II had begun.

The outbreak of war in Europe, coupled with additional warnings from the NACA committees and from Lindbergh about American preparedness, triggered support for a third research center. British, French, and German military planes were reportedly faster and more able in combat than their American counterparts. Part of the reason, according to experts, was the European emphasis on liquid-cooled engines that yielded benefits in speed and high altitude operations. In the United States, the country's large size had led to the development of air-cooled engines that were more suited to longer ranges and fuel efficiency. Moreover, according to Lindbergh, the NACA's earlier agreement to leave engine development to the manufacturers left the country with inadequate national research facilities for aircraft engines. Congress quickly responded, and an "Aircraft Engine Research Laboratory" was set up near the municipal airport in Cleveland, Ohio. This third new facility in the midwest gave the NACA a geographical balance, and the location also put it in a region that already had significant ties to the powerplant industry.

The site at Moffett field became Ames Aeronautical Laboratory in 1940, in honor of Dr. Joseph Ames, charter member of the NACA and its longtime chairman. The "Cleveland laboratory" remained just that until 1948, when it was renamed the Lewis Flight Propulsion Laboratory, in memory of its veteran director of research, George Lewis. Key personnel for both new laboratories came from Langley, and the two junior labs tended to defer to Langley for some time. By 1945, after several years of managing their own wartime projects, the Ames and Cleveland laboratories felt less like adolescents and more like peers of Langley. The NACA, like NASA after it, became a family of labs, but with strong individual rivalries.

Drag reduction studies on the Brewster XF2A-1 Buffalo influenced many later military fighters.

In the meantime, requirements of national security took priority. One significant project undertaken on the eve of World War II demonstrated the sort of work at Langley that had a major influence on aircraft design for years afterward. During 1938, the Navy became frustrated with the performance of a new fighter, the XF2A Brewster Buffalo. After the navy flew a plane to Langley, technicians set it up in the full scale tunnel for drag tests. It took only five days to uncover a series of small but negative aspects in the plane's design.

To the casual eye, the 250 MPH fighter with retractable gear appeared aerodynamically "clean." But the wind tunnel evaluations pinpointed many specific design aspects that created drag. The exhaust ports, gunsight, guns, and landing gear all protruded into the slipstream during flight; the accumulated drag effects hampered the plane's performance. By revamping these and other areas, the NACA reported a 10 percent increase in speed. Such a performance improvement, without raising engine power or reducing fuel efficiency, immediately caught the attention of other designers. Within the next two years, no fewer than 18 military prototypes went through the "cleanup" treatment given to the XF2A. Even though the Brewster Buffalo failed to win an outstanding combat record, others did, including the Grumman XF4F Wildcat, the Republic XP47 Thunderbolt, and the Chance Vought XF4N Corsair. The enhanced performance of these planes often represented the margin between victory and defeat in air combat. Moreover, specialists in the analysis of engine cooling and duct design later set the guidelines for inducing air into a postwar generation of jet engines.

The pace of war created personnel problems, especially when selective service began to claim qualified males after 1938. In the early years of the war, NACA personnel officers did considerable traveling each month to get deferments for employees working on national defense projects. Nonetheless, the NACA sometimes lost more employees than it was able to recruit. The issue was not resolved until early in 1944, when all eligible Langley employees were inducted into the Air Corps Enlisted Reserves, then put on inactive status under the exclusive management of NACA. The NACA draftees were given honorable discharges after Japan's surrender in 1945. The issue of the draft was not a threat to women, who made up about one-third of the entire staff by the end of the war. Although most of the female employees held traditional jobs as secretaries, increasing numbers held technical positions in the laboratories. Some did drafting and technical illustrating; some did strain-gauge measurements; others made up entire computing groups who worked through reams of figures pouring out of the various wind tunnels. A few held engineering posts. If women at Langley did not advance as rapidly in civil service as their male counterparts, most of the female employees later recalled that their treatment at the NACA was better than average when compared to other contemporary employers.

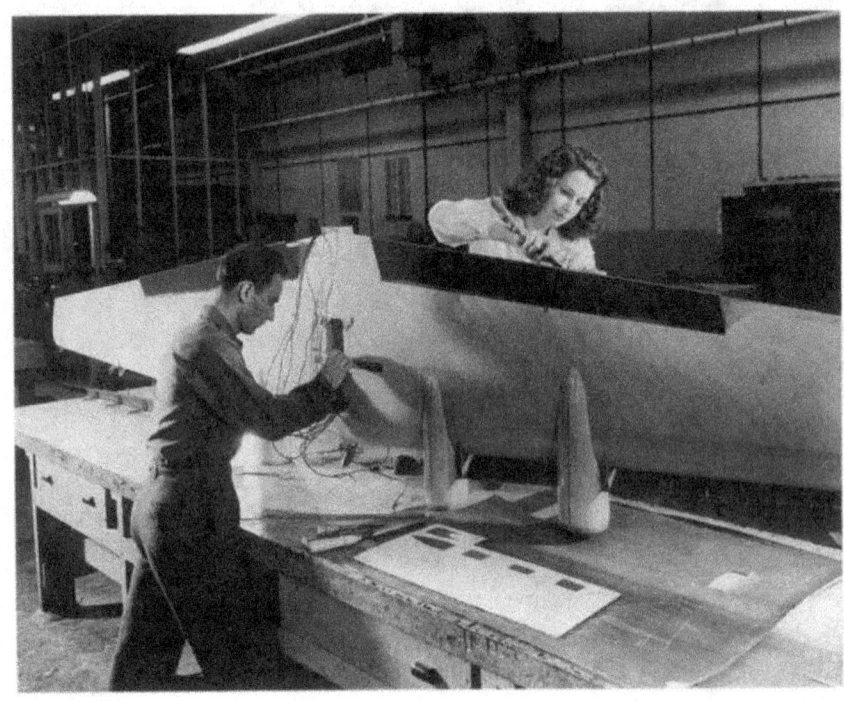

More women joined the NACA during World War II; technicians prepaired wind tunnel models, like this flying boast wing, for realistic tests.

Over the course of the war years, the NACA's relationship with industry went through a fundamental change. Since its inception, the agency refused to have an industry representative sit on the main committee, fearing that industry influence would make the NACA into a "consulting service." But the need to respond to industry goals in the emergency atmosphere of war led to a change in policy. The shift came in 1939, when George Mead became vice-chairman of the NACA and chairman of the Power Plants Committee. Mead had recently retired as a vice-president of the United Aircraft Corporation, and his position in the NACA, considering his high level corporate connections, represented a new trend. During the war, dozens of corporate representatives descended on Langley to observe and actually assist in testing. In the process, they forged additional direct links between the NACA and aeronautical industries.

Early in the war, extensive analysis of the Lockheed P-38 Lightning solved problems in high-speed dives.

Much of the wartime work involved refinement of manufacturers' designs, ranging from fighters through bombers like the B-29. Aircraft as large as the B-29 design were not tested as full sized planes, but considerable data was generated from models. During 1942, the B-29 design was thoroughly investigated in Langley's 8-foot high-speed tunnel, and Boeing engineers heaped praise on Langley technicians for their cooperation and the high quality of the data generated by the tests.

Despite the success of American warplanes, two of the major aeronautical trends of the era nearly escaped the NACA's attention. The agency endured much criticism in the postwar era for its apparent lapse in the development of jet propulsion and in the area of high-speed research leading to swept wings. America's rapid postwar progress in these fields suggest that there may have been a lapse of sorts, although not as total as many critics believed.

Rocketry

There was nothing in the original NACA charter that charged it with research in rocketry. Some of the NACA's personnel had a personal interest in rocketry, but most early developments in this field came from sophisticated amateur associations like the American Interplanetary Society. During World War II, governments suddenly became more interested in rocketry as a powerful new weapon.

The existence of organized groups like the VfR in Germany signaled the increasing fascination with modern rocketry in the 1930s, and there was frequent exchange of information among the VfR and other groups, like the British Interplanetary Society (1933) and the American Interplanetary Society (1930). Even Goddard occasionally had correspondence in the American Interplanetary Society's Bulletin, but he remained aloof from other American researchers, cautious about his results, and concerned about patent infringements. Because of Goddard's reticence, in contrast to the more visible personalities in the VfR, and because of the publicity given the German V-2 of the Second World War, the work of British, American, and other groups during the 1930s has been overshadowed. Their work, if not as spectacular as the V-2 project, nevertheless contributed to the growth of rocket technology in the prewar era and to the successful use of a variety of Allied rocket weapons in the Second World War. Although groups like the American Interplanetary Society (which became the American Rocket Society in 1934) succeeded in building and launching several small chemical rockets, much of their significance lay in their role as the source of a growing number of technical papers on rocket technologies.

But rocket development was complex and expensive. The cost and the difficulties of planning and organization meant that, sooner or later, the major work in rocket development would have to occur under the aegis of permanent government agencies and government-funded research bodies. In America, significant team research began in 1936 at the Guggenheim Aeronautical Laboratory, California Institute of Technology, or GALCIT. In 1939, this group received the first federal funding for rocket research, achieving special success in rockets to assist aircraft takeoff. The project was known as JATO, for jet-assisted takeoff, since the word "rocket" still carried negative overtones in many bureaucratic circles. JATO research led to substantial progress in a variety of rocket techniques, including both liquid and solid propellants. Work in solid propellants proved especially fortuitous for the United States; during the Second World War, American armed forces made wide use of the bazooka (an antitank rocket) as well as barrage rockets (launched from ground batteries or from ships) and high velocity air-to-surface missiles.

The most striking rocket advance, however, came from Germany. In the early 1930s the VfR attracted the attention of the German army, since armament restrictions introduced by the Treaty of Versailles had left the door open to rocket development. A military team began rocket research as a variation of long-range artillery. One of the chief assistants was a 22-year-old enthusiast from the VfR, Wernher von Braun, who joined the organization in October 1932. By December, the army rocket group had static-fired a liquid propellant rocket engine at the army's proving grounds near Kummersdorf, south of Berlin. During the next year it became evident that the test and research facilities at Kummersdorf would not be adequate for the scale of the hardware under development. A new location, shared jointly by the German army and air force, was developed at Peenemuende, a coastal area on the Baltic Sea. Starting with 80 researchers in 1936, there were nearly 5000 personnel at work by the time of the first launch of the awesome, long-range V-2 in 1942. Later in the war, with production in full swing, the work force swelled to about 18,000.

Having completed his doctorate in 1934 (on rocket combustion), von Braun became the leader of a formidable research and development team in rocket technology at Peenemuende. Like so many of his cohorts in original VfR projects, von Braun still harbored an intense interest in rocket development for manned space travel. Early in the V-2 development agenda, he began looking at the rocket in terms of its promise for space research as well as its military role, but found it prudent to adhere rigidly to the latter. Paradoxically, German success in the wartime V-2 program became a crucial legacy for postwar American space efforts.

Chapter 3

GOING SUPERSONIC (1945-1958)

On 1 October 1942, the Bell XP-59A, America's first jet plane, took to the air over a remote area of the California desert. There were no official NACA representatives present. The NACA, in fact, did not even know the aircraft existed, and the engine was based entirely on a top secret British design. After the war, the failure of the United States to develop jet engines, swept wing aircraft, and supersonic designs was generally blamed on the NACA. Critics argued that the NACA, as America's premier aeronautical establishment (one which presumably led the world in successful aviation technology) had somehow allowed leadership to slip to the British and the Germans during the late 1930s and during World War II.

In retrospect, the NACA record seems mixed. There were some areas, such as gas turbine technology, in which the United States clearly lagged, although NACA researchers had begun to investigate jet propulsion concepts. There were other areas, such as swept wing designs and supersonic aircraft, in which the NACA had made important forward steps. Unfortunately, the lack of advanced propulsion systems, such as jet engines, made such investigations academic exercises. The NACA's forward steps undeniably trailed the rapid strides made in Europe.

Jet Propulsion

During the 1930s, aircraft speeds of 300-350 MPH represented the norm and designers were already thinking about planes able to fly at 400-450 MPH. At such speeds, the prospect of gas turbine propulsion became compelling. With a piston engine, the efficiency of the propeller began to fall off at high speeds, and the propeller itself represented a significant drag factor. The problem was to obtain sufficient research and development funds for what seemed to be unusually exotic gas turbine power plants.

In England, RAF officer Frank Whittle doggedly pursued research on gas turbines through the 1930s, eventually acquiring some funding through a private investment banking firm after the British Air Ministry turned him down. Strong government support finally materialized on the eve of World War II, and the single-engine Gloster experimental jet fighter flew in the spring of 1941. English designers leaned more toward the centrifugal-flow jet engine, a comparatively uncomplicated gas-turbine design, and a pair of these power plants equipped the Gloster Meteor of 1944. Although Meteors entered RAF squadrons before the end of the war and shot down German V-1 flying bombs, the only jet fighter to fly in air-to-air combat came from Germany --the Me-262. Hans von Ohain, a researcher in applied physics and aerodynamics at the University of Gottingen, had unknowingly followed a course of investigation that paralleled Whittle's work and took out a German patent on a centrifugal engine in 1934. Research on gas turbine engines evolved from several other sources shortly thereafter, and the German Air Ministry, using funds from Hitler's rearmament program, earmarked more money for this research. Although a centrifugal type powered the world's first gas turbine aircraft flight by the He-178 in 1939, the axial-flow jet, more efficient and capable of greater thrust, was used in the Me-262 fighters that entered service in the autumn of 1944.

In America, the idea of jet propulsion had surfaced as early as 1923, when an engineer at the Bureau of Standards wrote a paper on the subject, which was published by the NACA. The paper came to a negative conclusion: fuel consumption would be excessive; compressor machinery would be too heavy; high temperatures and high pressures were major barriers. These were assumptions that subsequent studies and preliminary investigations seemed to substantiate into the 1930s. By the late 1930s, the Langley staff became interested in the idea of a form of jet propulsion to augment power for military planes for takeoff and during combat. In 1940, Eastman Jacobs and a small staff came up with a jet propulsion test bed they called the "Jeep." This was a ducted-fan system, using a piston engine power plant to combine the engine's heat and exhaust with added fuel injection for brief periods of added thrust, much like an afterburner. A test rig was in operation during the spring of 1942. By the summer, however, the Jeep had grown into something else --a research aircraft for transonic flight. With Eastman Jacobs again, a small team made design studies of a jet plane having the ducted fan system completely closed within the fuselage, similar to the Italian Caproni-Campini plane that flew in 1942. Although work on the Jeep and the jet plane design continued into 1943, these projects had already been overtaken by European developments.

During a tour to Britain in April 1941, General H. H. "Hap" Arnold, Chief of the U.S. Army Air Forces, was dumbfounded to learn about a British turbojet plane, the Gloster E28/39. The aircraft had already entered its final test phase and, in fact, made its first flight the following month. Fearing a German invasion, the British were willing to share the turbojet technology with America. That September, an Air Force Major, with a set of drawings manacled to his wrist, flew from London to Massachusetts, where General Electric went to work on an American copy of Whittle's turbojet. An engine, along with Whittle himself, followed. Development of the engine and design of the Bell XP-59 was so cloaked in secrecy that the NACA learned nothing about them until the summer of 1943. Moreover, design of the Lockheed XP-80, America's first operational jet fighter, was already under way.

General Arnold may have lost confidence in the NACA's potential for advanced research when he stumbled onto the British turbojet plane. It may be that British and American security requirements were so strict that the risks of sharing information with the civilian agency, where the risk of leaks was magnified, justified Arnold's decision to exclude the NACA. The answers were not clear. In any case, the significance of turbojet propulsion and rising speeds magnified the challenges of transonic aerodynamics. This was an area where the NACA had been at work for some years, though not without influence from overseas.

Shaping New Wings

As information on advanced aerodynamics began to trickle out of defeated Germany, American engineers were impressed. Photographs of some of the startling German aircraft, like the bat-like Me-163 rocket powered interceptor and the improbable Junkers JU-287 jet bomber, with its forward swept wings, prompted critics to ask why American designs appeared to lag behind the Germans. It seemed to be the story of the turbojet again. The vaunted NACA had let advanced American flight research fall precariously behind during the war. True, the effect of wartime German research made an impact on postwar American development of swept wings, leading to high performance jet bombers like the Boeing B-47 and the North American F-86 jet fighter. It is also the case that American engineers, including NACA personnel, had already made independent progress along the same design path when the German hardware and drawings were turned up at the end of World War II.

The North American F-86 Sabre featured swept wing and tail surfaces. The plane shown here was fitted with special instrumentation for transonic flight research conducted by the Ames Laboratory.

Like several other chapters in the story of high speed flight, the story began in Europe, where an international conference on high speed flight--the Volta Congress--met in Rome during October 1935. Among the participants was Adolf Busemann, a young German engineer from Lubeck. As a youngster, he had watched innumerable ships navigating Lubeck's harbor, each vessel moving within the V-shaped wake trailing back from the bow. As an aeronautical engineer, this image was a factor that led him to consider designing an airplane with swept wings. At supersonic speeds, the wings would function effectively inside the shock waves stretching back from the nose of an airplane at supersonic speeds. In the paper Busemann presented at the Rome conference, he analyzed this phenomenon and predicted that his "arrow wing" would have less drag than straight wings exposed to the shock waves.

There was polite discussion of Busemann's paper, but little else, since propeller-driven aircraft of the 1930s lacked the performance to merit serious consideration of such a radical design. Within a decade, the evolution of the turbojet dramatically changed the picture. In 1942, designers for the Messerschmitt firm, builders of the remarkable Me-262 jet fighter, realized the potential of swept wing aircraft and studied Busemann's paper more intently. Following promising wind tunnel tests, Messerschmitt had a swept wing research plane under development, but the war ended before the plane was finished.

In the United States, progress toward swept wing design proceeded independently of the Germans, although admittedly behind them. The American chapter of the swept wing story originated with Michael Gluhareff, a graduate of the Imperial Military Engineering College in Russia during World War I. He fled the Russian revolution and gained aeronautical engineering experience in Scandinavia. Gluhareff arrived in the United States in 1924 and joined the company of another Russian compatriot, Igor Sikorsky. By 1935, he was chief of design for Sikorsky Aircraft and eventually became a major figure in developing the first practical helicopter. In the meantime, Gluhareff became fascinated by the possibilities of low-aspect ratio tailless aircraft and built a series of flying models in the late 1930s. In a memo to Sikorsky in 1941, he described a possible pursuit-interceptor having a delta-shaped wing swept back at an angle of 56 degrees. The reason, he wrote, was to achieve "a considerable delay in the action (onset) of the compressibility effect. The general shape and form of the aircraft is, therefore, outstandingly adaptable for extremely high speeds."

Eventually, a wind tunnel model was built; initial tests were encouraging. But the Army declined to follow up due to several other unconventional projects already under way. Fortunately, a business associate of Gluhareff kept the concept alive by using the Dart design, as it was called, as the basis for an air-to-ground glide bomb in 1944. This time, the Army was intrigued and asked the NACA to evaluate the project. Thus, a balsa model of the Dart, along with some data, wound up on the desk of Robert T. Jones, a Langley aerodynamicist.

Jones was a bit of a maverick. A college dropout, he signed on as a mechanic for a barnstorming outfit known as the Marie Meyer Flying Circus. Jones became a self-taught aerodynamicist who couldn't find a job during the 1930s depression. He moved to Washington, D.C., and worked as an elevator operator in the Capitol. There he met a congressman who paid Jones to tutor him in physics and mathematics. Impressed by Jones's abilities, the legislator got him into a Works Projects Administration program that led to a job at Langley in 1934. With his innate intelligence and impressive intuitive abilities, Jones quickly moved ahead in the NACA hierarchy.

Studying Gluhareff's model, Jones soon realized that the lift and drag figures for the Dart were based on outmoded calculations for wings of high-aspect ratio. Using more recent theory for low-aspect ratio shapes, backed by some theoretical work done by Max Munk, Jones suddenly had a breakthrough. Within the shock cone created at supersonic speeds, he realized that the Dart's swept wing would remain free of shock waves at given speeds. The flow of air around the wings remained subsonic; compressibility effects would occur at higher Mach numbers than previously thought (Mach 1 equals the speed of sound; the designation is named after the Austrian physicist, Ernst Mach).

The concept of wings with subsonic sweep came to Jones in January 1945, and he eagerly discussed it with Air Force and NACA colleagues during the next few weeks. Finally, he was confident enough to make a formal statement to the NACA chieftains. On 5 March 1945, he wrote to the NACA's director of research, George W. Lewis. "I have recently made a theoretical analysis which indicates that a V-shaped wing traveling point foremost would be less affected by compressibility than other planforms," he explained. "In fact, if the angle of the V is kept small relative to the Mach angle, the lift and center of pressure remain the same at speeds both above and below the speed of sound."

So much for theory. Only testing would provide the data to make or break Jones's theory. Langley personnel went to work, fabricating two small models to see what would happen. Technicians mounted the first model on the wing of a P-51 Mustang. The plane's pilot took off and climbed to a safe altitude before nosing over into a high-speed dive towards the ground. In this attitude, the accelerated flow of air over the Mustang's wing was supersonic, and the instrumented model on the plane's wing began to generate useful data. For wind tunnel tests, the second model was truly a diminutive article, crafted of sheet steel by Jones and two other engineers. Langley's supersonic tunnel had a 9-inch throat, so the model had a 1.5-inch wingspan, in the shape of a delta. The promising test results, issued 11 May 1945, were released before Allied investigators in Europe had the opportunity to interview German aerodynamicists on delta shapes and swept wing developments.

Jones was already at work on variations of the delta, including his own version of the swept wing configuration. Late in June 1945, he published a summary of this work as NACA Technical Note Number 1033. Jones suggested that the proposed supersonic plane under development should have swept wings, but designers opted for a more conservative approach. Other design staffs were fascinated by the promise of swept wings especially after the appearance of the German aerodynamicists in America.

The Germans arrived courtesy of "Operation Paperclip," a high-level government plan to scoop up leading

German scientists and engineers during the closing months of World War II. Adolf Busemann eventually wound up at NACA's Langley laboratory, and scores of others joined Air Force, Army, and contractor staffs throughout the United States. Information from the research done by Robert Jones had begun to filter through the country's aeronautical community before the Germans arrived. Their presence, buttressed by the obvious progress represented by advanced German aircraft produced by 1945, bestowed the imprimatur of proof to swept wing configurations. At Boeing, designers at work on a new jet bomber tore up sketches for a conventional plane with straight wings and built the B-47 instead. With its long, swept wings, the B-47 launched Boeing into a remarkably successful family of swept wing bombers and jet airliners. At North American, a conventional jet fighter with straight wings, the XP-46, went through a dramatic metamorphosis, eventually taking to the air as the famed F-86 Sabre, a swept wing fighter that racked up an enviable combat record during the Korean conflict in the 1950s.

Nonetheless, America had been demonstrably lagging in jets and swept wing aircraft in 1945, and the NACA was the target of criticism from postwar Congressional and Air Force committees. It may have been that the NACA was not as bold as it might have been or that the agency was so caught up in immediate wartime improvements that crucial areas of basic research received short shrift. There were administrative changes to respond to these issues. In any case, as historian Alex Roland noted in his study of the NACA, Model Research (1985), its shortcomings "should not be allowed to mask its real significant contributions to American aerial victory in World War II." Moreover, the NACA's postwar achievements in supersonic research and rapid transition into astronautics reflected a new vigor and momentum.

The Sonic Barrier

During World War II, the increasing speeds of fighter aircraft began to create new problems. The Lockheed P-38 Lightning, for example, could exceed 500 MPH in a dive. In 1941, a Lockheed test pilot died when shock waves from the plane's wings (where the air flow over the wings reached 700 MPH) created turbulence that tore away the horizontal stabilizer, sending the plane into a fatal plunge. From wind tunnel tests, researchers knew something about the shock waves occurring at Mach 1, the speed of sound. The phenomenon was obviously attended by danger. Pilots and aerodynamicists alike muttered about the threatening dimensions of what came to be called the sound barrier.

Researchers faced a dilemma. In wind tunnels, with models exposed to near-sonic velocities, shock waves began bouncing from the tunnel walls, the "choking" phenomenon, resulting in questionable data. In the meantime, high speed combat maneuvers brought additional reports of control loss due to turbulence and, in several cases, crashes involving planes whose tails had wrenched loose in a dive. Since data from wind tunnels remained unreliable, researchers proposed a new breed of research plane to probe the sound barrier. Two of the leaders were Ezra Kotcher, a civilian on the Air Force payroll, and John Stack, on the NACA staff at Langley.

By 1944, John Stack and his NACA research team proposed a jet powered aircraft, a conservative, safe approach to high speed flight tests. Kotcher's group wanted a rocket engine which was more dangerous, with explosive fuels aboard, but more likely to achieve the high velocity to reach the speed of sound. The Air Force had the funds, so Stack and his colleagues agreed. The next problem involved design and construction of the rocket plane.

Eventually, the contract went to Bell Aircraft Corporation in Buffalo, New York. The company had a reputation for unusual designs, including the first American jet, the XP-59A Airacomet. The designer was Robert J. Woods, who had worked with John Stack at Langley in the 1920s before he joined Bell Aircraft. Woods had close contacts with the NACA as well as the Air Force. During a casual visit to Kotcher's office at Wright Field, Woods agreed to design a research plane capable of reaching 800 MPH at an altitude of 35,000 feet. Woods then called his boss, Lawrence Bell, to break the news. "What have you done?" Bell lamented, only half in jest.

The Bell design team worked closely with the Air Force and the NACA. This was the first time that the Langley staff had been involved in the initial design and construction of a complex research plane. Even with the Air Force bearing the cost and sharing the research load, this sort of collaboration marked a significant departure in NACA procedures. For the most part, design issues were amicably resolved, although some questions caused heated exchanges. The wing design was one such controversy.

There was general agreement that the wings would be thinner than normal in order to delay the formation of shock waves. In conventional designs, this was expressed as a numerical figure (usually between 12 to 15) which was the ratio of the wing's thickness to its chord. One group of NACA researchers advocated a 10 percent wing for the new plane, while others argued for an 8 percent thickness in order to forestall the effect of shock waves even more. One of Langley's resident experts on wing design finally made a thorough analysis of the issue and advised the 8 percent thickness as the most promising to achieve supersonic speed. As the design of the plane progressed, Bell's engineers came up with a plane that measured only 31 feet long with a wingspan of just 28 feet. Stresses on the remarkably short wing were estimated at twice the levels for high performance fighters of the day. Fortunately, Bell's designers realized that thickening the aluminum skin of the wings would result in a robust structure. Consequently, the skin thickness at the wing root measured .5 inch compared to .10-inch thick wing skin on a conventional fighter.

Research at Langley influenced other aspects of the design. Realizing that turbulence from the wing might create control problems around the tail, John Stack advised Bell to place the horizontal stabilizer on the fin, above the turbulent flow. He also recommended a stabilizer that was thinner than the wing, ensuring that shock waves would not form on the wing and tail at the same time, thereby improving the pilot's control over the accelerating aircraft. In making these decisions, the design team recognized that not much was known about the flight speeds for which the plane was intended. On the other hand, there was some interesting aerodynamic information available on the .50 caliber bullet, so the fuselage shape was keyed to ballistics data from this unlikely source. The cockpit was installed under a canopy that matched the rounded contours of the fuselage, since a conventional design atop the fuselage created too much drag.

The engine was one of the few really exotic aspects of the supersonic plane. Jet engines under development fell far short of the required thrust to reach Mach 1, forcing designers to consider rocket engines, a radical new technology for that time. The original engine candidate came from a small Northrop design for a flying wing. The propellants, red fuming nitric acid and aniline, ignited spontaneously when mixed. Curious about this volatile combination, some Bell engineers obtained some samples, put the stuff in a pair of bottles taped together, found some isolated rocks outside the plant, and tossed the bottles into them. They were aghast at the fierce eruption that followed. Considering the consequences to the plane and its pilot in case of a landing accident or a fuel leak, a different propulsion system seemed imperative. They settled on a rocket engine supplied by an outfit aptly named Reaction Motors, Incorporated. The engine burned a mixture of alcohol and distilled water along with liquid oxygen to produce a thrust of 1500 pounds from each of four thrust chambers. Due to limited propellant capacity of the research plane, the design team decided to use a Boeing B-29 Superfortress to carry it to about 25,000 feet. After dropping from the B-29 bomb bay, the pilot would ignite the rocket engine for a high-speed dash; with all its fuel consumed, the plane would have to glide earthward and make a deadstick landing. By this time, the plane was designated the XS-1, for Experimental Sonic 1, soon shortened to X-1 by those associated with it.

Early in 1946, flight trials began. The rocket engine was not ready, so the test crew moved into temporary quarters at Pinecastle Field, near Orlando, Florida. The X-1, painted a bright orange for high visibility, was carried aloft for a series of drop tests. By autumn, the X-1 was transferred to a remote air base in California's Mojave Desert--Muroc Army Air Field, familiarly known as Muroc,1 after a small settlement on the edge of Rogers Dry Lake. This was the Air Force flight test center, an area of 300 square miles of desolation in the California desert northwest of Los Angeles. Originating as an Air Force bombing and gunnery range, Muroc was a suitably remote location; the concrete-hard lake bed was highly suited for experimental testing. Test aircraft not infrequently made emergency landings, and the barren miles of Rogers Dry Lake allowed these unscheduled approaches from almost any direction. This austere, almost surrealistic desert setting made an appropriate environment for a growing roster of exotic planes based there in the postwar years.

The X-1 arrived under a cloud of gloom from overseas. The British had also been developing a plane to pierce the sound barrier, the de Havilland D.H. 108 Swallow, a swept wing, jet propelled, tailless airplane. Geoffrey, a son of the firm's founder, died during a high-speed test of the sleek aircraft in September 1946. The barrier was deadly.

Through the end of 1946 and into the autumn of 1947, one test flight after another took the X-1 to higher speeds, past Mach .85, the region where statistics on subsonic flight more or less faded away. On the one hand, the X-1 test crew felt increasing confidence that their plane could successfully make the historic run. On the other hand, NACA engineers like Walt Williams grudgingly admitted "a very lonely feeling as we began to run out of data."

The Air Force and the NACA put considerable trust in the piloting skills of Captain Charles "Chuck" Yeager, a World War II fighter ace. During the test sequences, he learned to keep his exuberance under control and to acquire a thorough knowledge of the X-1's quirks. On the morning of 14 October 1947, the day of the supersonic dash, Yeager's aggressive spirit helped him overcome the discomfort of two broken ribs, legacy of a horseback accident a few days earlier. A close friend helped the wincing Yeager into the cramped cockpit, then slipped him a length of broom handle so that he could secure the safety latch with his left hand, since the broken ribs on his right side made it too painful to use his right hand. The latch secure, Yeager reported he was ready to go. At 20,000 feet above the desert, the X-1 dropped away from the B-29.

Yeager fired up the four rocket chambers and shot upwards to 42,000 feet. Leveling off, he shut down two of the chambers while making a final check of the plane's readiness. Already flying at high speed, Yeager fired a third chamber and watched the instruments jump as buffeting occurred. Then the flight smoothed out; needles danced ahead as the X-1 went supersonic. Far below, test personnel heard a loud sonic boom slap across the desert. The large data gap mentioned by Walt Williams had just been filled in.

Ongoing Tests

A need for high-speed wind tunnel tests still existed. In the 7 x 10-foot tunnel at Langley, technicians built a hump in the test section; as the air stream accelerated over the hump, models could be tested at Mach 1.2 before the "choking" phenomenon occurred. A research program came up with the idea of absorbing the shock waves by means of longitudinal openings, or slots, in the test section. The slotted-throat tunnel became a milestone in wind tunnel evolution, permitting a full spectrum of transonic flow studies. In another high-speed test program, Langley used rocket-propelled models, launching them from a new test facility at Wallops Island, north of Langley on the Virginia coast. This became the Pilotless Aircraft Research Division (PARD), established in the autumn of 1945. During the next few years, PARD used rocket boosters to make high-speed tests on a variety of models representing new planes under development. These included most of the subsonic and supersonic aircraft flown by the armed services during the decades after World War II. In the 1960s, PARD facilities supported the Mercury, Gemini, and Apollo programs as well.

As full-sized aircraft took to the air, new problems inevitably cropped up. Researchers soon realized that a sharp increase in drag occurred in the transonic region. Slow acceleration through this phase of flight consumed precious fuel and also created control problems. At Langley, Richard T. Whitcomb became immersed in the problem of transonic drag. In the course of his analysis, Whitcomb developed a hunch that the section of an airplane where the fuselage joined the wing was a key to the issue. After listening to some comments by Adolph Busemann on airflow characteristics in the transonic regime, Whitcomb hit upon the answer to the drag problem- -the concept of the area rule.

Essentially, the area rule postulated that the cross-section of an airplane should remain reasonably constant from nose to tail, minimizing disturbance of the air flow and drag. But the juncture of the wing root to the fuselage of a typical plane represented a sudden increase in the cross-sectional area, creating the drag that produced the problems encountered in transonic flight. Whitcomb's solution was to compensate for this added wing area by reducing the area of the fuselage. The result was the "wasp-waisted" look, often called the "Coke bottle" fuselage. Almost immediately, it proved its value. A new fighter, Convair's XF-102, was designed as a supersonic combat plane but repeatedly frustrated the efforts of test pilots and aerodynamicists to achieve its design speed. Rebuilt with an area rule fuselage, the XF-102 sped through the transonic region like a champion; the Coke bottle fuselage became a feature on many high performance aircraft of the era: the F-106 Delta Dart (successor to the F-102), Grumman F-11, the Convair B-58 Hustler bomber, and others.

This Group portrait displays typical high-speed research aircraft that made headlines at Muroc Flight Center in the 1950s. The Bell X-1A (lower left) had much the same configuration as the earlier X-1. Joining the X-1A were (clockwise): the Douglas D-558-I Skystreak; Convair XF92-A, Bell X-5 with variable sweepback wings, Douglas D-558-II Skyrocket; Northrop X-4; and (center) the Douglas X-3

A succession of X aircraft, designed primarily for flight experiments, populated the skies above Muroc in a continuous cycle of research and development (R&D). Two more X-1 aircraft were ordered by the Air Force, followed by the X-1A and the X-1B, which investigated thermal problems at high speeds. The Navy used the Muroc flight test area for the subsonic jet-powered Douglas Skystreak, accumulating air-load measurements unobtainable in early postwar wind tunnels. The Skystreak was followed by the Douglas Skyrocket, a swept wing research jet (later equipped with a rocket engine that would surpass twice the speed of sound for the first time in 1953). The Douglas X-3, which fell short of expectation for further flight research in the Mach 2 range, nevertheless yielded important design insights on the phenomenon of inertial coupling (solving a control problem for the North American F-100 Super Sabre), the structural use of titanium (incorporated in the X-15 and other subsequent supersonic fighter designs), and data applied in the design of the Lockheed F-104 Starfighter. The NACA kept involved throughout these programs. In a number of ways, the X aircraft contributed substantially to the solution of a variety of high-speed flight conundrums and enhanced the design of future jet airliners, establishing a record of consistent progress aside from the speed records that so fascinated the public.

This photo taken from below the Grumman F-11 Navy fighter illustrates the way in which the area-ruled fuselage was adapted to production aircraft.

Although much of the NACA's work in this era had to do with military aviation, a good number of aerodynamic lessons were applicable to nonmilitary research planes and to civil aircraft. In the late 1950s, the Air Force began developing the North American XB-70, an unusually complex bomber capable of sustained supersonic flight over long distances. As a high-altitude strategic bomber, the B-70 was eventually displaced by ballistic missiles and a tactical shift to the idea of low-altitude strikes to avoid enemy radars and antiaircraft rockets. The Air Force and the NACA continued to fly the plane for research. Despite the loss of one of the two prototypes in a tragic midair collision involving a chase plane, the remaining XB-70 generated considerable data on long- range, high-altitude supersonic operations. This data was useful in designing new generations of jet transports operating in the transonic region, as well as advanced military aircraft.

Helicopters, introduced into limited combat service at the end of World War II, entered both military and civilian service in the postwar era. The value of helicopters in medical evacuation was demonstrated time and again in Korea, and a variety of helicopter operations proliferated in the late 1950s. The NACA flight-tested new designs to help define handling qualities. Using wind tunnel experience, researchers also developed a series of special helicopter airfoil sections, and a rotor test tower aided research in many other areas.

As usual, NACA researchers also pursued a multifaceted R&D program touching many other aspects of flight. In one project, the NACA installed velocity-gravity-altitude recorders in aircraft flown in all parts of the world. The object was to acquire information about atmospheric turbulence and gusts so that designers could make allowances for such perturbations. At Langley, a Landing Loads Track Facility went into operation, using a hydraulically propelled unit that subjected landing gear to the stresses of repeated landings in a variety of conditions. Another test facility studied techniques in designing pressurized fuselage structures to avoid failures. In the mid-1950s, a rash of such failures in the world's first operational jet airliner, the British-built de Havilland Comet, dramatized the rationale for this kind of testing.

All of this postwar aeronautical activity received respectful and enthusiastic attention from press and public. Although the phenomenon of flight continued to enjoy extensive press coverage, events in the late 1950s suddenly caused aviation to share the limelight with space flight.

Enter Astronautics

Among the legacies of World War II was a glittering array of new technologies spawned by the massive military effort. Atomic energy, radar, antibiotics, radio telemetry, the computer, the large rocket, and the jet engine seemed destined to shape the world's destiny in the next three decades and heavily influence the rest of the century. The world's political order had been drastically altered by the war. Much of Europe and Asia were in ashes. Old empires had crumbled; national economies were tottering perilously. On opposite sides of the world stood the United States and the Soviet Union, newly made into superpowers. It soon became apparent that they would test each other's mettle many times before a balance of power stabilized. And each nation moved quickly to exploit the new technologies.

The atomic bomb was the most obvious and most immediately threatening technological change from World War II. Both superpowers sought the best strategic systems that could deliver the bomb across the intercontinental distances that separated them. Jet-powered bombers were an obvious extension of the wartime B-17 and B-29, and both nations began putting them into service. The intercontinental rocket held great theoretical promise, but seemed much further down the technological road. Atomic bombs were bulky and heavy; a rocket to lift such a payload would be enormous in size and expense. The Soviet Union doggedly went ahead with attempts to build such rockets. The American military temporarily settled upon jet aircraft and smaller research and battlefield rockets. The Army imported Wernher von Braun and the German engineers who had created the wartime V-2 rockets and set them to overseeing the refurbishing and launching of V-2s at White Sands, New Mexico. The von Braun team was later transferred to Redstone Arsenal, Huntsville, Alabama, where it formed the core of the Army Ballistic Missile Agency (ABMA). With its contractor the Jet Propulsion Laboratory (JPL), the Army developed a series of battlefield missiles known as Corporal, Sergeant, and Redstone. The Navy designed and built the Viking research rockets. The freshly independent Air Force started a family of cruise missiles, from the jet Bomarc and Matador battlefield missiles to Snark and the ambitious rocket-propelled Navaho, which were intended as intercontinental weapons.

By 1951 progress on a thermonuclear bomb of smaller dimensions revived interest in the long-range ballistic missile. Two months before President Truman announced that the United States would develop the thermonuclear bomb, the Air Force contracted with Consolidated Vultee Aircraft Corporation (later Convair) to resume study, and then to develop, the Atlas intercontinental ballistic missile, a project that had been dormant for

38

four years. During the next four years three intermediate range missiles; the Army's Jupiter, the Navy's Polaris, and the Air Force's Thor; and a second generation ICBM, the Air Force's Titan, had been added to the list of American rocket projects. All were accorded top national priority. Fiscal 1953 saw the Department of Defense (DoD) for the first time spend more than $1 million on missile research, development, and procurement. Fiscal 1957 saw the amount go over the $1 billion mark.

By the mid-1950s NACA had modern research facilities that had cost a total of $300 million, and a staff totaling 7200. Against the background of the "Cold War" between the U.S. and the U.S.S.R. and the national priority given to military rocketry, the NACA's sophisticated facilities inevitably became involved. With each passing year it was enlarging its missile research in proportion to the old mission of aerodynamic research. Major NACA contributions to the military missile programs came in 1955-1957. Materials research led by Robert R. Gilruth at Langley confirmed ablation as a means of controlling the intense heat generated by warheads and other bodies reentering the Earth's atmosphere; H. Julian Allen at Ames demonstrated the blunt-body shape as the most effective design for reentering bodies; and Alfred J. Eggers at Ames did significant work on the mechanics of ballistic reentry.

The mid-1950s saw America's infant space program burgeoning with promise and projects. As part of the U.S. participation in the forthcoming International Geophysical Year (IGY), it was proposed to launch a small satellite into orbit around the Earth. After a spirited design competition between the National Academy of Sciences-Navy proposal (Vanguard) and the ABMA-JPL candidate (Explorer), the Navy design was chosen in September 1955 as not interfering with the high-priority military missile programs, since it would use a new booster based on the Viking research rocket, and having a better tracking system and more scientific growth potential. By 1957 Vanguard was readying its first test vehicles for firing. The U.S.S.R. had also announced it would have an IGY satellite; the space race was extending beyond boosters and payloads to issues of national prestige.

On the military front, space activity was almost bewildering. The missiles were moving toward the critical flight-test phase. Satellite ideas were proliferating, though mostly on a sub-rosa planning basis; after Sputnik these would become Tiros, weather satellite; Transit, navigation satellite; Pioneer lunar probes; Discoverer research satellites; Samos, reconnaissance satellite; Midas, missile early-warning satellite. Payload size and weight were constant problems in all these concepts, with the limited thrust of the early rocket engines. Here the rapid advances in solid-state electronics came to the rescue by reducing volume and weight; with new techniques such as printed circuitry and transistors, the design engineers could achieve new levels of miniaturization of equipment. Even so, heavier payloads were obviously in the offing; more powerful engines had to be developed. So design was begun for several larger engines, topped by the monster F-1 engine, intended to produce eight times the power of the engines that lifted the Atlas, Thor, and Jupiter missiles.

All this activity, however, was still on the drawing board, work bench, or test stand on 4 October 1957, when the "beep, beep" signal from Sputnik 1 was heard around the world. The Soviet Union had orbited the world's first man-made satellite.

The American public's response was swift and widespread. It seemed equally compounded of alarm and chagrin. American certainty that the nation was always number one in technology had been rudely shattered. Not only had the Russians been first, but Sputnik 1 weighed an impressive 183 pounds against Vanguard's intended start at 3 pounds and working up to 22 pounds in later satellites. In a cold war environment, the contrast suggested undefined but ominous military implications.

Fuel for such apprehensions added up rapidly. Less than a month after Sputnik 1 the Russians launched Sputnik 2, weighing a hefty 1100 pounds and carrying a dog as passenger. President Eisenhower, trying to dampen the growing concern, assured the public of our as yet undemonstrated progress and denied there was any military threat in the Soviet space achievements. As a counter, the White House announced the impending launch in December of the first Vanguard test vehicle capable of orbit and belatedly authorized von Braun's Army research team in Huntsville to try to launch their Explorer-Jupiter combination. But pressures for dramatic action gathered rapidly. The media ballyhooed the carefully qualified announcement on Vanguard into great expectations of America's vindication. On 25 November Lyndon B. Johnson, Senate majority leader, chaired the first meeting of the Preparedness Investigation Subcommittee of the Senate Armed Services Committee. The hearings would review the whole spectrum of American defense and space programs.

A ball of fire and flying debris mark the explosive failure of the first American attempt to launch a satellite on Vanguard, 6 December 1957.

Still the toboggan careened downhill. On 6 December 1957, the much-touted Vanguard test vehicle rose about 3 feet from the launch platform, shuddered, and collapsed in flames. Its tiny 3-pound payload broke away and lay at the edge of the inferno, beeping impotently.

A moment of triumph with the announcement that Explorer I has become the first American satellite to orbit the Earth. Here a duplicate Explorer is held aloft by (left to right) William H. Pickering of JPL, James A. van Allen of the State University of Iowa, and Wernher von Braun of the ABMA.

Clouds of gloom deepened into the new year. Then, finally, a small rift. On 31 January 1958, an American satellite at last went into orbit. Not Vanguard but the ABMA-JPL Explorer had redeemed American honor. True, the payload weighed only 2 pounds against the 1100 of Sputnik 2. But there was a scientific first; an experiment aboard the satellite reported mysterious saturation of its radiation counters at 594 miles altitude. Professor James A. van Allen, the scientist who had built the experiment, thought this suggested the existence of a dense belt of radiation around the Earth at that altitude. American confidence perked up again on 17 March when Vanguard 1 joined Explorer 1 in orbit.

Meanwhile, in these same tense months, both consensus and competition had been forming on the political front; consensus that an augmented national space program was essential; competition as to who would run such a program, in what form, with what priorities. The DoD, with its component military services, was an obvious front runner; the Atomic Energy Commission, already working with nuclear warheads and nuclear propulsion, had some congressional support, particularly in the Joint Committee on Atomic Energy; and there was NACA.

NACA had devoted more and more of its facilities, budget, and expertise to missile research in the mid- and late 1950s. Under the skillful leadership of James H. Doolittle, chairman, and Hugh L. Dryden, director, the strong NACA research team had come up with a solid, long-term, scientifically based proposal for a blend of aeronautic and space research. Its concept for manned spaceflight, for example, envisioned a ballistic spacecraft with a blunt reentry shape, backed by a world-encircling tracking system, and equipped with dual automatic and manual controls that would enable the astronaut gradually to take over more and more of the flying of his spacecraft. Also NACA offered reassuring experience of long, close working relationships with the military services in solving their research problems, while at the same time translating the research into civil applications. But NACA's greatest political asset was its peaceful, research-oriented image. President Eisenhower and Senator Johnson and others in Congress were united in wanting above all to avoid projecting cold war tensions into the new arena of outer space.

By March 1958 the consensus in Washington had jelled. The administration position (largely credited to James R. Killian in the new post of president's special assistant for science and technology), the findings of Johnson's Senate subcommittee, and the NACA proposal converged. America needed a national space program. The military component would of course be under DOD But a civil component, lodged in a new agency, technologically and scientifically based, would pick up certain of the existing space projects and forge an expanded program of space exploration in close concert with the military. All these concepts fed into draft legislation. On 2 April 1958, the administration bill for establishing a national aeronautics and space agency was submitted to Congress; both houses had already established select space committees; debate ensued; a number of refinements were introduced; and on 29 July 1958 President Eisenhower signed into law P.L. 85-568, the National Aeronautics and Space Act of 1958.

The act established a broad charter for civilian aeronautical and space research with unique requirements for dissemination of information, absorbed the existing NACA into the new organization as its nucleus, and empowered broad transfers from other government programs. The National Aeronautics and Space Administration came into being on 1 October 1958.

All this made for a very busy spring and summer for the people in the small NACA Headquarters in Washington. Once the general outlines of the new organization were clear, both a space program and a new organization had to be charted. In April, Dryden brought Abe Silverstein, assistant director of the Lewis Laboratory, to Washington to head the program planning. Ira Abbott, NACA assistant director for aerodynamic research, headed a committee to plan the new organization. In August President Eisenhower nominated T. Keith Glennan, president of Case Institute of Technology and former commissioner of the Atomic Energy Commission, to be the first administrator of the new organization, NASA, and Dryden to be deputy administrator. Quickly confirmed by the Senate, they were sworn in on 19 August. Glennan reviewed the planning efforts and approved most. Talks with the Advanced Research Projects Agency identified the military space programs that were space science-oriented and were obvious transfers to the new agency. Plans were formulated for building a new center for space science research, satellite development, flight operations, and tracking. A site was chosen, nearly 500 acres of the Department of Agriculture's research center in Beltsville, Maryland. The Robert H. Goddard Space Flight Center (named for America's rocket pioneer) was dedicated in March 1961.

Chapter 4

ON THE FRINGES OF SPACE (1958-1964)

On 1 October 1958, the 170 people in Headquarters gathered in the courtyard of their building, the Dolly Madison House, to hear Glennan proclaim the end of the 43-year-old NACA and the beginning of NASA. The 8000 people, three laboratories (now renamed research centers) and two stations, with a total facilities value of $300 million and an annual budget of $100 million were transferred intact to NASA. On the same day, by executive order the President transferred to NASA: Project Vanguard and its 150-person staff and remaining budget from the Naval Research Laboratory; lunar probes from the Army; lunar probes and rocket engine programs, including the F-1, from the Air Force; and a total of over $100 million of unexpended funds. NASA immediately delegated operational control of these projects back to the DoD agencies while it put its own house in order.

There followed an intense two-year period of organization, build up, fill in, planning, and general catch up. Only one week after NASA was formed, Glennan gave the go ahead to Project Mercury, America's first manned spaceflight program. The Space Task Group, headed by Robert R. Gilruth, was established at Langley to get the job done. The new programs brought into the organization were slowly integrated into the NACA nucleus. Many space-minded specialists were drawn into NASA, attracted by the exciting new vistas. Long-range planning was accelerated; the first NASA 10-year plan was presented to Congress in February 1960. It called for an expanding program on a broad front: manned flight (first orbital, then circumlunar); scientific satellites to measure radiation and other features of the near-space environment; lunar probes to measure the lunar space environment and to photograph the Moon; planetary probes to measure and to photograph Mars and Venus; weather satellites to improve our knowledge of Earth's broad weather patterns; continued aeronautical research; and development of larger launch vehicles for lifting heavier payloads. The cost of the program was expected to vary between $1 billion and $1.5 billion per year over the 10-year period.

Towards Hypersonic Flight

As NASA labored to get itself organized in the new field of astronautics, its traditional work in aeronautics experienced notable success. When the NACA set up the Muroc Flight Test Unit in 1948, Walter C. Williams began a decade of administration that saw many dramatic changes in the shapes and speeds of aircraft. The Muroc site won independence from Langley when it became the High-Speed Flight Station in 1954. Williams always argued for even more independence in the form of laboratory status, which would not only boost morale but also give the station greater prestige and autonomy. When NASA was created and the existing NACA labs were renamed as centers, old Muroc hands witnessed another change in names, becoming the NASA Flight Research Center (FRC) in 1959. Williams had to savor the change in names from a distance, since he already had been posted back to Langley as operations director for Project Mercury. But he could take pleasure at FRC's rapid growth and fame during the early 1960s, due largely to the test program for the X-15, a remarkably productive aircraft. After winning major headlines at the start of its flight tests, the X-15's success became eclipsed by NASA's space program. This was ironic, since the X-15 contributed heavily to research in spaceflight as well as to high-speed aircraft research.

The X-15 series were thoroughbreds, capable of speeds up to Mach 6.72 (4534 MPH) at altitudes up to

354,200 feet (67 miles). There was a familiar European thread in the design's genesis. In the late 1930s and during World War II, German scientists Eugen Sanger and Irene Bredt developed studies for a rocket plane that could be boosted to an Earth orbit and then glide back to land. The idea reshaped American thinking about hypersonic vehicles. "Professor Sanger's pioneering studies of long-range rocket-propelled aircraft had a strong influence on the thinking which led to initiation of the X-15 program," NACA researcher John Becker wrote. "Until the Sanger and Bredt paper became available to us after the war we had thought of hypersonic flight only as a domain for missiles...." A series of subsequent studies in America "provided the background from which the X-15 proposal emerged."

Momentum for such a plane gathered in 1951, when Robert Woods, the X-1 veteran from Bell Aircraft, proposed a Mach 5 research plane. Woods argued his case in the prestigious NACA Committee on Aerodynamics, of which he was a member. The NACA Committee took no formal action, but independent projects got underway at Ames, Langley, and FRC (Edwards). By 1954, the NACA accepted the hypersonic aircraft proposal as a major commitment. By autumn of that year, the NACA realized it lacked funds to support the idea and joined forces with the Air Force and Navy; a Memorandum of Understanding gave the NACA technical control of the effort, including flight testing and test reports. There was an undertone of military necessity in the Memorandum, which declared that "accomplishment of this project is a matter of national urgency." The specifications and configurations circulated among potential bidders followed a pattern originally developed by a Langley team led by John Becker. "The proposals that we got back looked pretty much like the one we had put in," he recalled. The NACA had certainly come a long way from testing aircraft designed and built by others. The earlier X-1 was something of a transition, involving Bell and NACA engineers. Although the NACA in essence bootstrapped Air Force and Navy funds for the X-15, it was very much a NACA idea and design from start to finish. In many ways, the X-15 program represented a shift to the research, development, and management functions that characterized the NASA organization soon to come.

The X-15 streaks across the western United States on a test run. Capable of flying at 6.7 times the speed of sound at altitudes of over 350,000 feet, the X-15 helped advance many aeronautical and space flight systems.

In the fall of 1955, North American emerged as the winning contractor. Aside from building the plane, the NACA and armed services soon realized that they had also had to develop other elements of a new system to support flight tests of the exotic X-15. The program called for fabrication of three research planes and a powerful new rocket engine to power them. The engine, a Thiokol XLR-99, had to be "man-rated" for repeated flights in the piloted rocket plane. For pilot training and familiarization, it was necessary to design and build a motion simulator and associated analog computer equipment. Before making a 10-to 12-minute mission in the X-15, pilots eventually spent 8 to 10 hours practicing each moment of the test flight. Due to the extreme altitudes planned for X-15 missions, technicians needed to develop a unique, full-pressure flight suit. Finally, planners had to lay out a special aerodynamic test range to monitor the X-15 as the plane streaked back to Edwards Air Force Base for its landing.

The test range, officially labeled the High Altitude Continuous Tracking Radar Range, became known as the "High Range." The increased speeds of research planes meant that customary air-to-ground communications at the test field were outmoded. The High Range stretched 485 miles from Wendover Air Force Base in Utah to Edwards in California. A trio of tracking stations along the route were built and equipped with advanced radar and telemetry, recording equipment, and consoles for monitoring the X-15. All the tracking stations passed real-time data to each other as the X-15 sped down the High Range. With its experience in the acquisition of in-flight data, NACA expertise in setting up the High Range was invaluable. Following the X-15 program, the High Range continued to be a continuing asset to flight testing of succeeding generations of aircraft.

The first X-15 arrived in the autumn of 1958, although powered flight tests did not start until September of 1959. In contrast to the secrecy surrounding the P-59 and the X-1, the X-15 program was a high-visibility media event. In the wake of Sputnik, anything that seemed to redeem America's tarnished prestige in the "space race" automatically occupied center stage. Journalists flocked to Edwards for photos and interviews; Hollywood cranked out a hackneyed film about terse, steely-eyed test pilots and the rocket-powered ships they flew. When the Mercury, Gemini, and Apollo programs began, the journalists migrated to hotter headlines in Florida. The X-15, meanwhile, moved into the most productive phase of its program, contributing to astronautics as well as aeronautics.

Between 1959 and 1968, the trio of X-15 aircraft completed 199 test flights. The fallout was far-reaching in numerous crucial areas, such as hypersonic aerodynamics and in structures. During a test series to investigate high-temperature phenomena in hypersonic flight, temperatures on the skin soared to 1300º F, so that large sections of the aircraft glowed a cherry-red color. The X-15's survival encouraged extensive use of comparatively exotic alloys, like titanium and Inconel-X, leading to machining and production techniques that became standard in the aerospace industry. Although the cockpit was pressurized, the chance of accidental loss of pressurization in the near-space environment where the X-15 flew prompted development of the first practical full-pressure suit for pilot protection in space. The X-15 was the first to use reaction controls for attitude control in space; reentry techniques and related technology also contributed to the space program, and even earth sciences experiments were carried out by the X-15 in some of its flights.

The high-speed, high-altitude X-15, like the X-1, might be remembered as the epitome of an era, although the NACA/NASA research activities, as usual, continued along many paths. For example, in the course of studies for supersonic cruise aircraft, two different trends of study began to emerge: a multi-mission combat plane operating at both high and low speeds, and configurations for a supersonic transport.

The Grumman F-14 Tomcat, with wings swept back for high-speed flight, was a legacy of variable geometry studies (photo courtesy of Grumman Aerospace Corporation).

The multi-mission plane idea took shape as a combat aircraft capable of sustained high speeds at high altitudes, as well as high speeds "down on the deck." This meant swept wings, which also decreased controllability and combat load at takeoff--unless the wings could be pivoted forward during takeoff and landing and swept back during flight. Test articles from wartime German experiments again pointed the way, and the Bell X-5 provided additional data during the early 1950s. The British also had a variable-sweep concept plane called the Swallow, which underwent extensive testing at Langley. The NASA contribution in this development included variable in-flight sweeping of the wings and the decision to locate the pivot points outboard on the wings rather than pivot the wings on the centerline, solving a serious instability problem. All of this eventually led to the TFX program, which became the F-111. It was a long and controversial program but the success of the variable geometry wing on the F-111 and the Navy's Grumman F-14 Tomcat owed much to NASA experimental work. The process of refining Mach 2 aircraft like these also led to profitable studies involving air inlets, exhaust nozzles, and overall drag reduction --factors that the aerospace industry applied to the new stable of Mach 2 combat planes of the following decades.

In addition to the dramatic high-speed military planes scrutinized by NASA, there was a slower plane with a truly unique ability: it could take off and land vertically. A considerable degree of effort went into a series of aircraft with a tilt-wing layout, like the Boeing Vertol 76. Langley built and tested a scale free-flight model, which was followed by a full-sized aircraft with a gas-turbine propulsion system driving a pair of oversized propellers. Concurrently, a variety of different configurations went through a test program in small wind tunnels while very large models were tested in the big 40 x 80-foot tunnel at Ames. One result of this combined activity was a tri-service transport experimental program for the Army, Air Force, and Navy. Known as the XC-142A, a one-ninth scale model went through remote control flight tests in Langley's full scale tunnel. There were additional tests carried out with full-sized experimental configurations built by Bell and by Ryan; flight testing continued into the 1980s.

The work in high-speed combat planes paralleled growing interest in a supersonic transport. In 1959, a delegation from Langley briefed E. R. Quesada, head of the FAA, on the technical feasibility of a supersonic transport (SST). The NASA group advocated a variable geometry wing and an advanced, fan-jet propulsion system. The briefing, later published as NASA Technical Note D-423, "The Supersonic Transport: --A Technical Summary," analyzed structures, noise, runways and braking, traffic control, and other issues related to SST operations on a regular basis. An SST, the report concluded, was entirely feasible. The FAA concurred, and within a year, a joint program with NASA had allocated contracts for engineering component development. Eventually, the availability of advanced Air Force aircraft provided the opportunity to conduct flight experiments as well. The idea of commercial airliners flashing around the globe at supersonic speeds received press attention, but the biggest headlines went to even more sensational developments in space, where human beings were preparing for inaugural voyages.

The New Space Program

To conduct its space program, NASA obviously needed capabilities it did not have. To that end Glennan sought to acquire the successful Army team that had launched America's first satellite, the ABMA at Huntsville, Alabama, and its contractor, the JPL in Pasadena, California. The Army balked at losing the Huntsville group, claiming it was indispensable to the Army's military rocket program. Glennan for the time being had to compromise: ABMA would work on NASA programs as requested. The Army grudgingly gave up JPL. On 3 December 1958, an executive order transferred, effective 31 December, the government-owned plant of JPL and the Army contract with the California Institute of Technology, under which JPL was staffed and operated. Glennan renewed his bid for ABMA in 1959; protracted Army resistance was finally overcome and on 15 March 1960 ABMA's 4000-person Development Operations Division, headed by Wernher von Braun, was transferred to NASA along with the big Saturn booster project.

As the 10-year plan took shape and the capability grew, there were many other gaps to be filled. NASA was going to be markedly different from NACA in two important ways. First, it was going to be operational as well as do research. So, it would not only design and build launch vehicles and satellites but it would launch them, operate them, track them, acquire data from them, and interpret the data. Second, it would do the greater part of its work by contract rather than in-house as NACA had done. The first of these required tracking sites in many countries around the world, as well as construction of facilities: antennae, telemetry equipment, computers, radio and landline communications networks, and so on. The second required the development of a larger and more sophisticated contracting operation than NACA had needed. In the first years, NASA leaned heavily on the DoD procurement system.

The problem of launch vehicles occupied much attention in these first years. A family of existing and future launch vehicles had to be structured for the kinds of missions and spacecraft enumerated in the plan. In addition to the existing Redstone, Thor, and Atlas vehicles, NASA would develop:

- Scout, a low-budget solid-propellant booster that could put small payloads in orbit;

- Centaur, a liquid-hydrogen-fueled upper stage, transferred from DoD, that promised higher thrust and bigger payloads for lunar and planetary missions;

- Saturn, which was expected to be flying in 1963 (with the proper upper stages it would put upwards of 46,000 pounds in Earth orbit);

- Nova, several times the size of Saturn, to be started later in the decade for the more ambitious manned lunar flights anticipated in the 1970s.

In addition, work could continue with the Atomic Energy Commission on the difficult but enormously promising nuclear-propelled upper stage, Nerva, and on the SNAP family of long-life electric power producers.

As much as larger boosters were needed, an even more immediate problem was how to improve the reliability of existing boosters. By December 1959 the United States had attempted 37 satellite launches; less than one-third attained orbit. Electrical components, valves, turbopumps, welds, materials, structures --virtually everything that went into the intricate mechanism called a booster-- had to be redesigned or strengthened or improved to withstand the stresses of launch. A new order of perfection in manufacturing and assembly had to be instilled in workers and managers. Rigorous, repeated testing had to verify each component, then subassembly, then total vehicle. That bugaboo of the engineering profession, constant fiddling and changing in search of perfection, had to be constrained in the interest of reliability. And since the existing vehicles were DoD products, NASA had to persuade DoD to enforce these rigorous standards on its contractors.

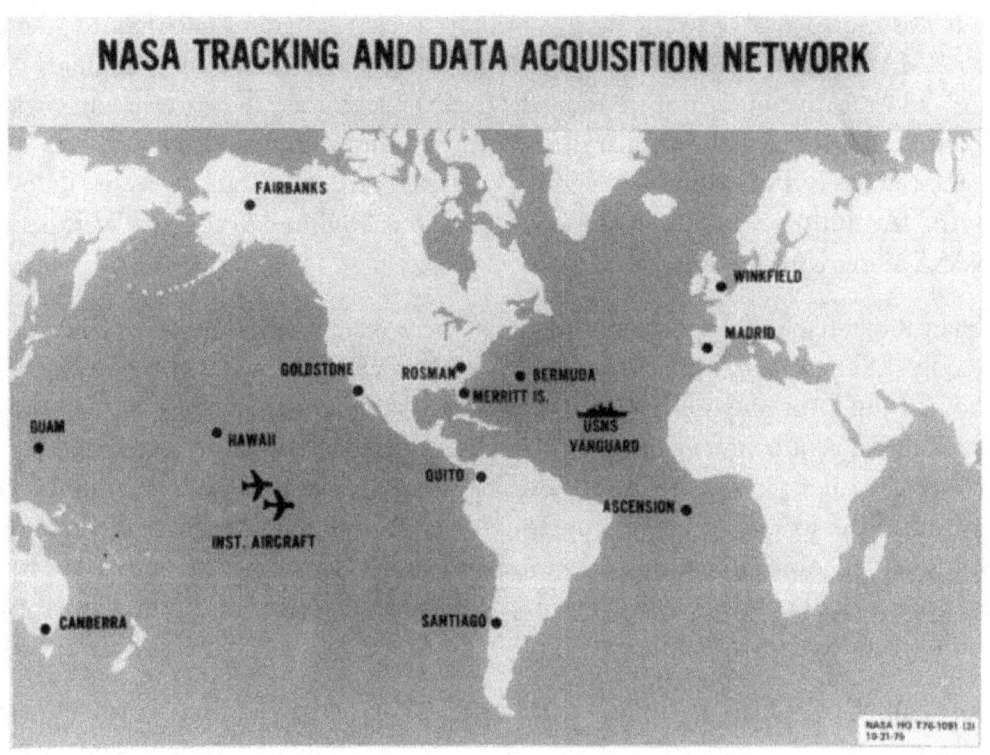

The worldwide satellite tracking network, 1975

That was only one of the areas in which close coordination between NASA and DoD was essential and effective. In manned spaceflight, for example, there were essentially four approaches to putting man into space:

- the research airplane--the Air Force and NASA were already well into this program, leading to the X-15;

- the ballistic vehicle--NASA's Project Mercury embodied this approach, with Air Force launch vehicles and DoD support throughout;

- the boostglider--the Air Force had inaugurated the DynaSoar project (later renamed the X-20) in

48

November 1957. A manned glider would be boosted into shallow Earth orbit, bounce in and out of the top of the atmosphere for part or all of a revolution of the planet, and land like an airplane. In May 1958 NACA had agreed to help with the technical side of the project. NASA continued that support;

- the lifting body--a bathtub-like shape proposed by Alfred J. Eggers of Ames Laboratory which, as a reentry shape, would be midway between an airplane configuration and the ballistic shape, developing moderate lift during reentry and landing like an airplane. This approach would be deferred for a few years before being explored by the Air Force and NASA.

In the communications satellites area DoD had its Courier program, a low-altitude, militarily-secure communications satellite; it also had Advent, intended to be put into equatorial synchronous orbit by the Atlas Centaur booster to provide global communications for the military. NASA had a passive communications satellite, Echo, a 98-meter inflatable sphere from which to bounce radar signals as a limited communications relay and, over a period of time and with accurate tracking, to plot the variations in air density at the top of the atmosphere by following the vagaries of its orbit. It had been agreed that NASA would leave active communications satellites (those that picked up, amplified, and rebroadcast radio signals from one point on Earth to another) to DoD. But this did not answer for long. By 1960 the American Telephone and Telegraph Company (AT&T) was asking NASA to launch its low-level, active communications satellite, Telstar. NASA also had another proposal for medium-altitude (roughly 11,125-mile orbit) communications satellites.

The AT&T proposal raised a fundamental problem: would industry develop communications satellites entirely with its own money or would the government fund such research? NASA sought and received presidential approval to go both ways--to provide reimbursable launches to industry and to do its own communications satellite research. First there was Relay, the medium-altitude repeater satellite. Beyond lay the imaginative proposal from Hughes Aircraft Company for Syncom, a synchronous-orbit satellite that would fly at 21,753- mile altitude, where distance, gravity, and velocity combined to place a satellite permanently over the same spot on Earth. By virtue of the lofty orbit, three of these satellites could cover the entire planet and require only a handful of ground stations.

By the time of the presidential election of 1960 the worst pangs of reorganization, redefinition, and planning were over. Programs were meshing with each other; contracting for large projects was becoming routine; the initial absorption of DoD programs had been completed; and a viable organization was in business.

There were operational bright spots as well. True, launch vehicles were still fickle and unpredictable; 7 out of 17 launches failed in 1959. But finally in August 1959, NASA launched its first satellite that functioned in all respects (Explorer 6). Pioneer 5, launched on 11 March 1960 and intended to explore interplanetary space between Earth and Venus, communicated out to a new distance record, 22 million miles. The first of the prototype weather satellites, Tiros 1, launched on 1 April 1960, produced 22,500 photos of Earth's weather. Echo 1, the first passive communications satellite, was launched 12 August 1960, inflated in orbit, and provided a passive target for bouncing long-range communications from one point on Earth to another. Perhaps as important, millions of people saw the moving pinpoint of light in the night sky and were awed by the experience.

In late 1960 politics bemused the space program. Although not a direct campaign issue in the presidential campaign, the space program found little reassurance of its priority as an expensive new item in the federal budget. After John F. Kennedy was narrowly elected, the uncertainty deepened. Jerome B. Wiesner, the President-elect's science adviser, chaired a committee which produced a report both critical of the space program's progress to date and skeptical of its future. Who would be the new administrator? What, if any, priority would the fledgling space program have in a new, on-record hostile administration?

Then, once again, challenge and response. On 12 April 1961, Soviet Cosmonaut Yuri Gagarin rode Vostok 1 into a 187 x 108 mile orbit of the Earth. After one orbit he reentered the atmosphere and landed safely. A human had flown in space. Gagarin joined that elite pantheon of individuals who were the first to do the undoable--Wright brothers, Lindbergh, now Gagarin. There was faint consolation on 6 May 1961, when Mercury essayed its first manned spaceflight. Astronaut Alan B. Shepard, Jr., rode a Redstone booster in his Freedom 7 Mercury spacecraft for a 15-minute suborbital flight and was picked out of the water some 300 miles downrange. Success, yes; a good beginning, yes. But Gagarin had flown around the Earth, some 24,800 miles against Shepard's 300. His Vostok weighed 10,428 pounds in orbit, contrasting with Mercury's 2,100 pounds in suborbit. Gagarin had had about 89 minutes in weightlessness, the mysterious zerogravity condition that had supplanted the sound barrier as the great unknown. Shepard experienced 5 minutes of weightlessness. By any unit of measure, clearly the United States was still behind, especially in the indispensable prerequisite of rocket power. As the new President had said, gloomily: "We are behind...the news will be worse before it is better, and it will be some time before we catch up." The public reaction was less emphatic than after Sputnik 1 but congressional concern was strong. Robert C. Seamans, Jr., NASA's associate administrator and general manager, was hard put to restrain Congress from forcing more money on NASA than could be effectively used.

NASA's seven original astronauts were all experienced test pilots. Posed in front of a Convair F-106, they are (left to right): Scott Carpenter, Gordon Cooper, John Glenn, Virgil Grissom, Walter Schirra, Alan Shepard, and Donald Slayton.

President Kennedy was especially concerned. His inaugural address in January had rung with an eloquent promise of bold new initiatives that would "get this country moving again." The succeeding three months had been distinguished by crushing setbacks --the Bay of Pigs invasion fiasco and the Gagarin flight. As one of several searches for new initiatives, the President asked his Vice President, Lyndon B. Johnson, to head a study of what would be required in the space program to convincingly surpass the Soviets. Johnson, the only senior White House figure in the new administration with prior commitment to the space program, found strong support waiting in the wings. James E. Webb, new administrator of NASA, had an established reputation as an aggressive manager of large enterprises, both in industry and the Truman administration as director of the Bureau of the Budget and undersecretary of state. Backed by the seasoned technical judgment of Dryden, his deputy, and Seamans, his general manager, Webb moved vigorously to accelerate and expand the central elements of the NASA 10-year plan.

The largest single concept in that plan had been manned circumlunar flight. Now the question became: could this country rally quickly enough to beat the Soviets to that circumlunar goal? The considered technical estimate was "not for sure." But if we went one large step further and escalated the commitment to manned lunar landing and return, it became a new ball game. Both nations would have to design and construct a whole new family of boosters and spacecraft; this would be an equalizer in terms of challenge to both nations and the experts were confident that the depth and competence of the American government-industry-university team would prove superior. In this judgment they found a strong ally in the new secretary of defense, Robert S. McNamara.

But Webb and his advisers were not content with a one-shot objective. The goal, they said, was a major space advance on a broad front- manned spaceflight, yes, but also boosters, communications satellites, meteorological satellites, and planetary exploration.

This was the combined proposal presented to the Vice President and approved and transmitted by him to the President. It was the best new initiative the President had seen. So it was that on 25 May 1961 the President stood before a joint session of Congress and proposed a historic national goal:

> Now it is time to take longer strides-time for a great new American enterprise-time for this nation to take a clearly leading role in space achievement, which in many ways may hold the key to our future on earth . . . . I believe that this nation should commit itself to achieving the goal, before this decade is out, of landing a man on the moon and returning him safely to the earth. No single space project in this period will be more impressive to mankind, or more important for the long-range exploration of space; and none will be so difficult or expensive to accomplish.

> The President correctly assessed the national mood. Editorial support was widespread. Congressional debate was perfunctory, given the size of the commitment. The decision to land an American on the Moon was endorsed virtually without dissent.

The Lunar Commitment

NASA was exhilarated but awed. Dryden had returned from a White House meeting to tell his staff that "this man" (Webb) had sold the President on landing a man on the Moon. Gilruth, immersed in what seemed to be big enough problems in the relatively modest Project Mercury, was temporarily aghast. But the die was cast. The nation had accepted the challenge to its largest technological enterprise, dwarfing even the wartime Manhattan Project for developing the atomic bomb and the postwar crash development of strategic missiles.

The blank check was there; the way to use it was far from clear. Since 1958, studies had been underway on a circumlunar manned flight. Since 1959, George M. Low, head of the manned spaceflight office in Headquarters, had ramrodded a series of progressively more detailed studies on the requirements for a manned landing on the Moon. Those studies had established a broad confidence that no major technological or scientific breakthroughs were needed to get a man to the Moon or even to land and return him. But there were some operational unknowns; the blank check caused them suddenly to loom larger. The assumption had been that one simply built a big enough booster, flew directly to the Moon, landed a large vehicle, and returned some part of it directly to Earth. But there were wide scientific disagreements as to the nature of the lunar surface. Was it solid "ground," strong enough to support such a load? Or was it many feet of dust, in which a spacecraft would disappear without a trace? Or was it something in between? There were operational problems: could the crew and ground control possibly handle the enormous peak of work that would bunch together in the landing phase of a direct-ascent mission? The alternative seemed to be that one boosted pieces of a lunar vehicle into Earth orbit, assembled and refueled them there, and took off for a direct landing on the Moon. This too was fraught with hazards: could payloads rendezvous in Earth orbit? Could men assemble complex equipment in the demanding environment of space? Could such operations as refueling with volatile fuels --hazardous enough on Earth-- be safely performed in space?

Some points were clear. The very massiveness of the effort would make this program different in kind from anything NASA had attempted. New organizational modes were essential; no one center could handle this program. A much stronger Headquarters team would be needed, coordinating the efforts of several centers and riding herd on an enormous mobilization of American industry and university effort.

Also, there were long leadtime problems that needed to be worked on irrespective of later decisions. One of these was three years under way --a big engine. Work on the 1.5 million-pound-thrust F-1 engine would be accelerated. Another was a navigation system; accurate vectoring of a spacecraft from Earth to a precise point on a rapidly moving Moon 230,000 miles away was a formidable problem in celestial mechanics. Therefore, the first large Apollo contract was let to the Massachusetts Institute of Technology and its Instrumentation Laboratory, headed by C. Stark Draper, to begin study of this inscrutable problem and to develop the requisite navigational system.

The basic spacecraft could be delineated --the one in which a crew would depart the Earth, travel to the Moon, and return. It should have a baggage car, a jettisonable service module housing its propulsion, expendable oxygen, and other equipment. The Space Task Group was hard at work on these with its left hand, while its main effort on Mercury went forward. That left hand had to be strengthened.

A whole new logistics system was needed; from factory to launch, everything had outstripped normal sizes and normal transportation. There would have to be new factories, mammoth test stands, huge launch complexes. Railroads and highways could not handle the larger components. Ship transportation seemed the only answer. A massive facility design and site location program had to begin even before the final configuration of the vehicle was decided. Limited in the facilities and construction area, NASA decided to call on the tested resource of the Army Corps of Engineers. It proved to be one of the wiser decisions in this hectic period.

As planning went forward in 1961 and 1962, order gradually emerged. A new concept for how to get to the Moon painfully surfaced: lunar-orbit rendezvous. A small group at Langley, headed by John C. Houbolt, had studied the trade-offs of direct ascent, Earth-orbit rendezvous, and other possibilities. They had been increasingly struck with the vehicle and fuel economics of this mission profile: after stabilizing in Earth orbit, a set of spacecraft went to orbit around the Moon, and, leaving the mother spacecraft in lunar orbit, dispatched a smaller craft to land on the lunar surface, reconnoiter, and rejoin the mother craft in lunar orbit for the return to Earth. Over a period of two years they refined their complex mathematics and argued their case. As time became critical for definition of the launch vehicle, they argued their case before one NASA audience after another. Finally Houbolt, in a bold move, went outside of "channels" and got the personal attention of Seamans. This was a decision of such importance to the total program that imposed decision was not enough; the major elements of NASA had to be won over and concur in the final technical judgment. Dismissed at first as risky and very literally "far out," lunar orbit rendezvous gradually won adherents. In July 1962 D. Brainerd Holmes, NASA director of manned spaceflight, briefed the House space committee on lunar orbit rendezvous, the chosen method of going to the Moon.

Once made, this decision permitted rapid definition of the Apollo spacecraft combination. Launch vehicle configuration had been arrived at seven months earlier. The objective would be to put a payload of nearly 300,000 pounds in Earth orbit and 100,000 pounds in orbit around the Moon. To do this required a three-stage vehicle, the first stage employing the F-1 engine in a cluster of five, to provide 7.5 million pounds of thrust at launch. The second stage would cluster five of a new 225,000-pound-thrust liquid hydrogen and liquid oxygen engine (the J-2). The third stage, powered by a single J-2 engine, would boost the Apollo three-man spacecraft out of Earth orbit and into the lunar gravitational field. At that point the residual three-spacecraft combination would take over: a command module housing the astronauts, a service module providing propulsion for maneuvers, and a two-man lunar module for landing on the Moon. The engine on the service module would ignite to slow the spacecraft enough to be captured into lunar orbit; the fragile lunar module would leave the mother craft and descend to land its two passengers on the Moon. After lunar reconnaissance, the astronauts would blast off in the top half of the lunar module to rejoin the mother craft in lunar orbit, and the service module would fire up for return to Earth.

A smaller launch vehicle, which would later be dubbed the Saturn IB, would be built first and used to test the Apollo spacecraft in Earth orbit. Even this partial fulfillment of the Apollo mission would require a first stage with 1.5 million pounds of thrust and a high-energy liquid oxygen-liquid hydrogen second stage.

Launches of the Saturn I (pictured) and the similar Saturn IB increased NASA's confidence in engines, boosters, and spacecraft, prepairing the way for eventual manned missions of the Apollo program.

The grand design was now complete. But in the articulating of it, vast gaps in experience and technology were revealed. At three critical points the master plan depended on successful rendezvous and docking of spacecraft. Although theoretically feasible, it had never been done and was not within the scope of Project Mercury. How could practical experience be gained with rendezvous and docking short of an intricate, hideously expensive, and possibly disastrous series of experiments with Apollo hardware? Men would, hopefully, land and walk upon the Moon. But could men and their equipment function in space outside the artificial and confining environment of their spacecraft? Other systems and other questions could be engineered to solution on Earth, but the ultimate questions here could only be answered in space. We had bitten off more than we could chew. Clearly something was needed between the first steps of Mercury and the grand design of Apollo. The gap was too great to jump when men's lives were at stake.

Mercury crew capsules crashed into the Atlantic while the Atlas and Apollo crew capsules splashed down into the Pacific, to be retrieved by helicopter. The Sikorsky UH-34D lost its struggle with Grissom's capsule, which sank after the astronaut scrambled out.

Even Mercury sometimes seemed a very big mouthful to chew. But slowly, stubborn problem after stubborn problem yielded. The second suborbital flight, Liberty Bell 7, was launched on 21 July 1961; its 16-minute flight went well, though on landing the hatch blew off prematurely and the spacecraft sank just after Astronaut Virgil I. Grissom was hoisted to safety in a rescue helicopter. In September the unmanned Mercury-Atlas combination was orbited successfully and landed where it was supposed to, east of Bermuda. On 29 November the final test flight took chimpanzee Enos on a two-orbit ride and landed him in good health. The system was qualified for manned orbital flight. And on 20 February 1962, Astronaut John H. Glenn, Jr., became the first American to orbit the Earth in space. Friendship 7 circled the Earth three times; Glenn flew parts of the last two orbits manually because of trouble with his autopilot.

The United States took its astronaut heroes to its heart with an enthusiasm that bewildered them and startled NASA. Their mail was enormous; hundreds of requests for personal appearances poured in. Glenn had a rainy parade in Washington and addressed a joint session of Congress. On 1 March four million people in New York showered confetti and ticker tape on him and fellow astronauts Shepard and Grissom. Nor was the event unnoticed by the competition. President Kennedy announced the day after the Glenn flight that Soviet Premier Nikita Khrushchev had congratulated the nation on its achievement and had suggested the two nations "could work together in the exploration of space." The results of this exchange were a series of talks between Dryden of NASA and Anatoliy A. Blagonravov of the Soviet Academy of Sciences. By the end of the year they had agreed to exchanges of meteorological and magnetic-field data and some communications experiments.

A big year for the young American space program, 1962. Two more Mercury flights, Carpenter for three orbits, then Schirra for six. The powerful Saturn I booster made two test flights, both successful. The first active communications satellite, Telstar I, was launched for AT&T by NASA; later NASA's own Relay communications satellite was orbited; and the first international satellite, Britain's Ariel I, was launched by NASA to take scientific measurements of the ionosphere. Mariner 2 became the first satellite to fly by another planet; on 14 December it passed within 21,380 miles of Venus and scanned the surface of that cloud-shrouded body, measuring its temperatures. Then it continued into orbit about the Sun, eventually setting a new communications distance record of 55.4 million miles. The fifth and sixth Tiros meteorological satellites were placed in orbit and continued to report the world's weather. So successful had Tiros been that the R&D program had quickly become semi-operational. The Weather Bureau was regularly integrating Tiros data into its operational forecasting and was busy planning a full scale weather satellite system which it would operate. The hard work on booster reliability began to pay off --18 successes to 9 failures or partial successes.

Not that all was sweetness and light. The Ranger, designed to photograph the Moon while falling to impact the lunar surface, was in deep trouble. A high-technology program at the edge of the state of the art, Ranger closed the year with five straight failures and another would come in 1963. JPL, the NASA agent; Hughes Aircraft Co., the contractor; and NASA Headquarters came under heavy pressure from Congress. Studies were made; a reorganization realigned JPL and contractor to firm commitment to the project; NASA dropped the science experiments; and the last three Ranger flights were spectacularly successful, providing close-in lunar photography that excelled the best telescopic detail of the Moon from Earth by 2000 times and dispelled many of the scare theories about the lunar surface.

As the dimensions of Apollo began to dawn on Congress and the scientific community, there were rumbles: Apollo would preempt too much of the scientific manpower of the nation; Apollo was an "other worldly" stunt, directed at the Moon instead of at pressing problems on Earth. Administrator Webb met both of these caveats with positive programs.

In acknowledgment of the drain on scientific manpower, Webb won White House support for a broad program by NASA to augment the scientific manpower pool. Thousands of fellowships were offered for graduate study in space-related disciplines, intended to replace or at least supplement the kinds of talent engulfed by the space program. Complementing the fellowships was an even more innovative program, government-financed buildings and facilities on university campuses for the new kinds of interdisciplinary training that the space program required.

From a modest beginning in 1962, by the end of the program in 1970 NASA had footed the bill for the graduate education of 5000 scientists and engineers at a cost of over $100 million, had spent some $32 million in construction of new laboratory facilities on 32 university campuses, and had given multidisciplinary grants to some 50 universities that totaled more than $50 million. The program marked a new direction in the government's recognition of its responsibility for impact of its program on the civilian economy and a new dimension of cooperation between the university and the government. In part as a result of these new capabilities in the universities, NASA contracts and grants for research by universities rose from $21 million in 1962 to $101 million in 1968. The NASA university program proved very effective: on the political side it reduced tensions between NASA and the scientific- engineering community; on the score of national technology capability it enlarged and focused a large segment of the research capabilities of the universities.

To refute the other charge --that Apollo would serve only its own ends and not the broader needs of the nation's economy-- Webb created the NASA technology utilization program in 1962. Its basic purpose was to identify and hold up to the light the many items of space technology that could be or had been adapted for uses in the civilian economy. By 1973 some 30,000 such uses had been identified and new ones were rolling in at the rate of 2000 a year.

But the program went beyond that. A concerted effort was made in every NASA center not only to identify possible transfers of space technology but to use NASA technical people and contractors to explore and even perform prototype research on promising applications. NASA publications described all these potential applications to researchers and industry; seven regional dissemination centers were established to work directly with industry on technical problems in the adaption of space technology; in 1973 some 2000 companies received direct help and another 57,000 queries were answered. New products ranged from quieter aircraft engines to microminiaturized and solidstate electronics that revolutionized TV sets, radios, and small electronic calculators. NASA's computer software programs enabled a wide range of manufacturers to test the life history of new systems; they could predict problems that could develop, how the systems would perform, how long they would last, and so on. Many other facets of the space program were important to the quality and sustenance of life for citizens of the United States and the world:

Communications. Within a decade the communications satellite proved to be a reliable, flexible, cost-effective addition to long- range communications. The Communications Satellite Corporation (Comsat) became a solid financial success, with 114,000 stockholders. As manager of the International Telecommunications Satellite Consortium (Intelsat), it shared access to the global satellite system with 82 other nations who had become members of the consortium. Its array of sophisticated Intelsat communications satellites bracketed the world from synchronous orbit. Before these satellites existed, the total capability for transoceanic telephone calls had been 500 circuits; in 1973 the Intelsat satellites alone offered more than 4000 transoceanic circuits. Real-time TV coverage of events anywhere in the world --whether Olympics, wars, or coronations-- had become commonplace in the world's living rooms. Satellite data transmission enabled industries to control far-flung production and inventories, airlines to have instantaneous coast-to-coast reservation systems, large banks to have nationwide data networks. This was only the beginning of the communications revolution. The next generation of communications satellite, Intelsat 5, started operations in 1976 with five times the capacity of its predecessor (Intelsat 4) and a life expectancy of 10 years in orbit. In 1976 the Maritime Administration embarked on a global ship-control system operated by means of satellites. Experiments with Applications Technology Satellites (ATS) would continue to refine the lifesaving biomedical communication network which links medical personnel and medical centers across the nation. Especially valuable to isolated and rural areas, the network would afford them real-time access to expert diagnosis and prescription of treatment.

Weather forecasting. Like its brother the communications satellite, the weather satellite had in less than a decade become an established friend of people around the world. Potentially disastrous hurricanes such as Camille in August 1969 and Agnes in June 1972 were spotted, tracked, and measured by the operational weather satellite network of the National Oceanic and Atmospheric Administration. The realtime knowledge of the storm's position, intensity, and track made possible accurate early warning and emergency evacuation that saved hundreds of lives and millions of dollars in property damage. Near-global rainfall maps were being produced by 1973 from data acquired by NASA's Nimbus 5. Not only did the heat-release information contained in such data markedly improve long-range weather forecasting, but the data were of immediate value in agriculture, flood control, and similar tasks. Ice-movement charts for the Arctic and Antarctic regions were extending shipping schedules in these areas by several months a year.

Medicine. NASA's experience in microminiaturized electronics and in protecting and monitoring the health of astronauts during spaceflight generated hundreds of medical devices and techniques that could save lives and improve health care. Multidisciplinary teams of space technicians and medical researchers were successful in developing long-duration heart pacers, for instance. Implanted in the patient's body but rechargeable from outside, the tiny pacer would regulate the heartbeat for decades without replacement, whereas the previous model required surgical replacement every two years. Space-derived automatic patient monitoring systems were being used in more and more hospitals. Tiny sensors on the patient's body would trigger an alarm when there was a significant change in temperature, heartbeat, blood pressure, or even in the oxygen- carbon dioxide levels in the blood --a signal of the onset of shock. For researchers living inside space simulators for long periods of time, the Ames Research Center developed an aspirin--sized transmitter pill. In general medical practice, the transmitter pill was swallowed by the patient; as it moved through the digestive system it radioed to the doctor diagnostic measurements of any of several kinds of deep body conditions such as temperature, stomach acid level, etc.

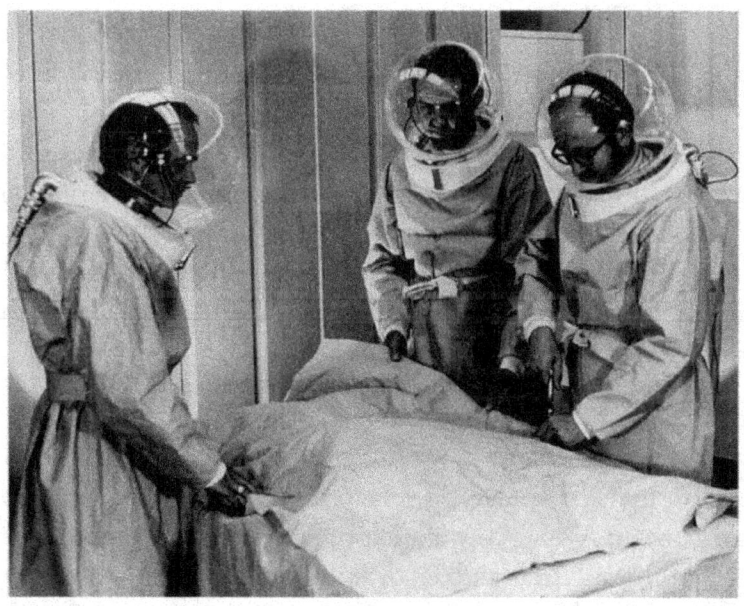

Laminar flow clean room and special clothing used at St. Luke's Hospital, Denver, in 1972 to lower risk of infection in hip joint replacements and other surgical procedures. Both the room and the clothing were based on space program experience and were developed under NASA contract by the Martin-Marietta Corporation.

Energy. The nation's stepped-up program of energy research that began in 1973 found NASA with broad experience and an existing program of research in devices that collect, store, transmit, and apply solar, nuclear, and chemical energy for production of mechanical and electrical power. Solar cells had produced the electric power for several generations of spacecraft; when arrays of them were experimentally mounted on houses they supplied as much as three-quarters of the energy needed to heat and cool the house. But solar cells were too expensive to be competitive with other systems; work was continuing on improving their efficiency and on new manufacturing techniques that would cut their cost in half. A long-standing problem with the efficient use of electrical energy has been the inability to store significant amounts of it for future use. NASA had done much work on developing more compact, higher storage capacity, longer-life batteries. Nickel-cadmium batteries developed for the space program were already in general use; they could be recharged in 6 to 20 minutes instead of the 16 to 24 hours required for conventional batteries. Silver-zinc batteries used in spacecraft were too expensive for commercial use, but their unique separator material could double the capacity of conventional nickel-zinc batteries. An extensive trial of this adaptation was begun with the fleet of Postal Service electric trucks. Batteries with 5 to 20 times the storage capacity of conventional mass-produced automobile batteries could have a wide range of uses: low-pollution automobile propulsion; storage of excess electrical power generated during low-demand hours and released at times of peak demand; emergencies; and other uses. Fuel cells had been developed by NASA to provide the longer duration Gemini and Apollo flights with electrical power; on Earth they could be used either for energy storage or energy conversion. One of the ingredients used in fuel cells was hydrogen; in this application hydrogen was broken down and combined with oxygen in a complex chemical process that produced water and electrical energy. But hydrogen is also a superb high-performance, low-pollutant fuel whose source is inexhaustible. Liquid hydrogen had propelled men to and from the Moon. With its years of work with hydrogen as a rocket fuel, NASA had more experience than anyone else in the production, transportation, storage, pumping, and use of hydrogen. One possible use of hydrogen was a compact, clean energy that could be transported into large urban areas. Many kinds of Earth-based power plants could burn hydrogen, alone or in various combinations, to produce energy with low pollution side effects.

Apollo Impact. The creation of NASA's university and technology transfer programs in the early 1960s could be considered a side effect of Apollo. There were others. All lunar reconnaissance programs had been impacted by Apollo. The latter part of Ranger had been reoriented; Surveyor, the first lunar soft-lander, was reconfigured to support Apollo. If Surveyor worked, it would provide on-the-lunar-surface photography plus televised digging in the surface of the Moon for a better sense of soil composition. The remaining problem for Apollo was the need for detailed mapping photography of the Moon. So by the end of 1963 a third program was initiated -- Lunar Orbiter, a state-of-the-art mapping satellite that would go into orbit around the Moon and photograph potential landing zones for Apollo.

The vexing questions of rendezvous and extravehicular activity still had to be answered. So on 3 January 1962 NASA announced a new manned spaceflight project, Gemini. Using the basic configuration of the Mercury capsule enlarged to hold a two-man crew, Gemini was to fit between Mercury and Apollo and provide early answers to assist the design work on Apollo. The launch vehicle would be the Titan II missile being developed by the Air Force. More powerful than Atlas and Titan I, it would have the thrust to put the larger spacecraft into Earth orbit. For a target vehicle with which Gemini could rendezvous, NASA chose the Air Force's Agena; launched by an Atlas, the second-stage Agena had a restartable engine that enabled it to have both passive and active roles. Gemini would be managed by the same Space Task Group that was operating Mercury; the project director would be James A. Chamberlin, an early advocate of an enlarged Mercury capsule.

Gemini began as a Mark II Mercury, a "quick and dirty" program. The only major engineering change aside from scale-up was to modularize the various electrical and control assemblies and place them outside the inner shell of the spacecraft to simplify maintenance. But perhaps not an engineer alive could have left it at that. After all, Gemini was supposed to bridge to Apollo. Here was a chance to try out ideas. If they worked, they would be available for Apollo. There was the paraglider, for example, that Francis Rogallo had been experimenting with at Langley. If that worked, Gemini could forget parachutes and water landings with half the Navy out there; with a paraglider Gemini could land routinely on land. The spacecraft should be designed to have more aerodynamic lift than Mercury, so the pilot could have more landing control; fuel cells (instead of batteries) with enough electric power to support longer duration flights; and fighter plane-type ejection seats for crew abort, to supersede the launch escape rocket that perched on top of Mercury.

All these innovations were cranked into the program, and contracts and subcontracts were let for their design and fabrication. Soon the monthly bills for Gemini were running far beyond what had been budgeted. In every area, it seemed, there were costly problems. The paraglider and ejection seats would not stabilize in flight; the fuel cell leaked; Titan II had longitudinal oscillations --the dreaded "pogo" effect-- too severe for manned flights; Agena had reconfiguration problems. Cost overruns had become severe by late 1962; by March 1963 they were critical. The original program cost of $350 million had zoomed to over $1 billion --$200 million higher than the figures Associate Administrator Seamans had used in Congress a few days before! Charles W. Mathews, the new program manager, cracked down. Flight schedules were stretched out; the paraglider gradually slid out of the program. By early 1964 most of the engineering problems were responding to treatment.

With the Mercury program and the spacecraft design role in Apollo, and now Gemini, it was clear that the Space Task Group needed a home of its own and some growing room. On 19 September 1961, Administrator Webb announced that a new Manned Spacecraft Center would be built on the outskirts of Houston. It would house the enlarged Space Task Group, now upgraded to a center, and would have operational control of all manned missions as well as be the developer of manned spacecraft. Water access to the Gulf of Mexico was provided by the ship channel to Galveston.

Water access played a role in all site selections for new Apollo facilities. The big Michoud Ordnance Plant outside New Orleans, where the 10-meter-diameter Saturn V first stage would be fabricated, was on the Mississippi River; the Mississippi Test Facility, with its huge test stands for static firing tests of the booster stages, was just off the Gulf of Mexico, in Pearl River County, Mississippi.

All this effort would come together at the launch site at Cape Canaveral, Florida, where NASA had a small Launch Operations Center, headed by Kurt H. Debus. NASA had been a tenant there, using Air Force launch facilities and tracking range. Now Apollo loomed. Apollo would require physical facilities much too large to fit on the crowded Cape. For safety's sake there would have to be large buffer zones of land around the launch pads; if a catastrophic accident occurred, where all stages of the huge launch vehicle exploded at once, the force of the detonation would approach that of a small atomic bomb. So NASA sought and received congressional approval to purchase over 111,000 acres of Merritt Island, just northwest of the Air Force facilities. Lying between the Banana River and the Atlantic, populated mostly by orange growers, Merritt Island had the requisite water access and safety factors.

Kennedy Space Center as it appeared in the mid-1960s. The 350-foot tall Saturn V launch vehicle has emerged from the cavernous Vehicle Assembly Building aboard its crawler and begun its stately processional to the launch complex three miles away.

Planners struggled through 1961 with a wide range of concepts and possibilities for the best launch system for Apollo, hampered by having only a gross knowledge of how the vehicle would be configured, what the missions would involve, and how frequent the launches would be. Finally on 21 July 1962 NASA announced its choice: the Advanced Saturn (later Saturn V) launch vehicle would be transported to the new Launch Operations Center on Merritt Island stage by stage; the stages would be erected and checked out in an enormous vehicle assembly building; the vehicle would be transported to one of the four launch pads several miles away by a huge tractor crawler. This system was a major departure from previous practice at the Cape; launch vehicles had usually been erected on the launch pad and checked out there. Under the new concept the vehicle would be on the launch pad for a much shorter time, allowing for a higher launch rate and better protection against weather and salt spray. As with the other new Apollo facilities, the Corps of Engineers would supervise the vast construction project.

The simultaneous building of facilities and hardware was going to take a great deal of money and a great many skilled people. The NASA budget, $966.7 million in fiscal 1961, was $1.825 billion in 1962. It hit $3.674 billion the next year and by 1964 was $5.1 billion. It would remain near that level for three more years. In personnel, NASA grew in those same years from 17,471 to 35,860. Of course this was small potatoes compared to the mushrooming contractor and university force where 90 percent of NASA's money was spent. When the Apollo production line peaked in 1967, more than 400,000 people were working on some aspect of Apollo.

Indeed, as the large bills began to come in, there was some wincing in the political system. President Kennedy wondered briefly if the goal was worth the cost; in 1963 Congress had its first real adversary debate on Apollo. Administrator Webb had to point out again and again that this was not a one-shot trip to the Moon but the building of a national space capability that would have many uses. He also needled congressmen with the fact that the Soviets were still ahead; in 1963 they were orbiting two-man spacecraft, flying a 129 mile orbit tandem mission, and orbiting an unmanned prototype of a new spacecraft. Support rallied. The Senate rejected an amendment that would have cut the fiscal 1964 space budget by $500 million. The speech that President Kennedy was driving through Dallas to deliver on that fateful 22 November 1963 would have defended the expenditures of the space program:

> This effort is expensive--but it pays its own way, for freedom and for America . . . . There is no longer any doubt about the strength and skill of American science, American industry, American education and the American free enterprise system. In short, our national space effort represents a great gain in, and a great resource of, our national strength.

As 1963 drew to a close, NASA could feel that it was on top of its job. The master plan for Apollo was drawn; the organization and the key people were in place. Mercury had ended with L. Gordon Cooper's 22-orbit flight, far beyond the design limits of the spacecraft. For those Americans old enough to have thrilled to Lindbergh's historic transatlantic flight 36 years earlier, it was awesome that in only 50 minutes more flight time, Cooper had flown 593,500 miles to Lindbergh's 3107. Of 13 NASA launches during the year, 11 were successful. In addition to improved performance from the established launch vehicles, *Saturn* I had another successful test flight, as did the troublesome Centaur. The *Syncom* 2 communications satellite achieved synchronous orbit and from that lofty perch transmitted voice and teletype communications between North America, South America, and Africa. The *Explorer* 18 scientific satellite sailed out in a long elliptical orbit to measure radiation most of the way to the Moon.

Chapter 5

TORTOISE BECOMES HARE (1964-1969)

As 1964 dawned, the worst of Gemini's troubles were behind. The spacecraft for the first flight was already at the Kennedy Space Center (Launch Operations Center, renamed in November 1963 by President Lyndon B. Johnson) being minutely checked out for the flight. Too minutely, too time-consumingly. Not until 8 April did *Gemini* I lift off unmanned into an orbit which confirmed the launch vehicle-spacecraft combination in the rigors of launch. The excessive checkout time of *Gemini* I generated a new procedure. Beginning with the next spacecraft, a contingent from the launch crew would work at the factory (McDonnell Douglas in St. Louis) to check out the spacecraft there. When it arrived at the Cape, it would be ready to be mated with its Titan II, have the pyrotechnics installed, and be launched. Only in this way could one hope to achieve the three-month launch cycle planned for Gemini.

The new system delayed the arrival of the second Gemini spacecraft at the Cape. There the curse set in. Once on the pad the spacecraft was struck by lightning, threatened by not one but two hurricanes, and forced to undergo check after check. And when launch day finally came in December, the engines ignited and then shut down. More rework. Finally on 19 January 1965, Gemini 2 rose from the launch pad on the tail of almost colorless flame from Titan II's hypergolic propellants, and in a 19-minute flight confirmed the readiness of a fully equipped Gemini spacecraft and the integrity of the heatshield during reentry. Gemini was man-rated.

The final test flight, a manned, three-orbit qualification flight, was conducted on 23 March without incident. Now the diversified flight program could continue. One program objective was to orbit men in space for at least the week that it would take an Apollo flight to go to the Moon, land, and return. Gemini 4 (3-7 June) stayed aloft four days; Gemini 5 (21-29 August) doubled that time and surpassed the Soviet long-duration record; Gemini7 (4-18 December) provided the clincher with 14 days (330 hours, 35 minutes). Of more lasting importance than the durability of the equipment was the encouraging medical news that no harmful effects were found from several weeks exposure to weightlessness. There were temporary effects, of course: heartbeat slowed down, blood tended to pool in the legs, the bones lost calcium, and other conditions appeared, but things seemed to stabilize after a few days in weightlessness and to return to normal after a few days back on Earth. So far there seemed to be no physiological time limit for humans living in space.

A crucial question for Apollo was whether the three rendezvous and docking maneuvers planned for every lunar flight were feasible. Gemini 3 made the tentative beginning by testing the new thruster rockets with short-burst firings that changed the height and shape of orbit, and one maneuver that for the first time shifted the plane of the flight path of a spacecraft. Gemini 4 tried to rejoin its discarded second-stage booster but faulty techniques burned up too much maneuvering fuel and the pursuit had to be abandoned --a valuable lesson; back to the computers for better techniques! Gemini 5 tested out the techniques and verified the performance of the rendezvous radar and rendezvous display in the cockpit.

Then came what is still referred to by NASA control room people with pride but also with slight shudders as "Gemini 76." The original mission plan called for a target Agena stage to be placed in orbit and for Gemini to launch in pursuit of it. But the Agena fell short of orbit and splashed into the Atlantic. The Gemini spacecraft suddenly had no mission. Round-the-clock debate and recomputation produced a seemingly bizarre solution, which within three days of the Agena failure was approved by Administrator Webb and President Johnson: remove the Gemini 6 spacecraft-launch vehicle combination intact from the launch pad and store it carefully to preserve the integrity of checkout; erect Gemini 7 on the launch pad, check it out and launch it; bring Gemini 6 out and launch it to rendezvous with the long-duration Gemini 7. It happened. Gemini 7 was launched 4 December 1965; Gemini 6 was back on the pad for launch by 12 December. On launch day the engines ignited, burned for four seconds, and shut off automatically when a trouble light lit up. On top of the fueled booster Astronaut Walter M. Schirra, Jr., sat with his hand on the lanyard of the ejection seat while the control checked out the condition of the fueled booster. But the potential bomb did not explode. On 15 December Gemini 6 lifted off to join its sister ship in orbit. On his fourth orbit Schirra caught up to Gemini 7 and maneuvered to within 33 feet; in subsequent maneuvers he moved to within six inches. Rendezvous was feasible; was docking?

On 16 March 1966, Gemini 8 on its third orbit docked with its Agena target. Docking too was feasible, though in this case not for long. Less than half an hour after docking for an intended full night in the docked position, the two spacecraft unaccountably began to spin, faster and faster. Astronaut Neil A. Armstrong could not stabilize the joined spacecraft, so he fired his Gemini thrusters to undock and maneuver away from the Agena. Still he could not control his single spacecraft with the thrusters; lives seemed in jeopardy. Finally he fired the reentry rockets, which did the job. By then ground control had figured out that one thruster had stuck in the firing position. Armstrong made an emergency landing off Okinawa. Despite hardware problems, docking had been established as feasible.

Rendezvous was new and difficult, so experimentation continued. Gemini 9 (3-6 June 1966) tried three kinds of rendezvous maneuvers with a special target stage as its passive partner, but docking was not possible because the shroud covering the target's docking mechanism had not separated. The shroud did not prevent simulation of an Apollo lunar orbit rendezvous. Gemini 10 (18-21 July 1966) did dock with its Agena target and used the powerful Agena engine to soar to a height of 474 miles, the highest in space man had ventured. It rendezvoused with the derelict Agena left in orbit by Gemini 8 four months earlier, using only optical methods and thereby demonstrating the feasibility of rendezvous with passive satellites for purpose of repairing them. On the next flight Gemini 11 caught up with its target in its first orbit, demonstrating the possibility of quick rendezvous if necessary for rescue or other reasons. Each astronaut practiced docking twice. Using Agena propulsion, they rocketed out to 850 miles above the Earth, another record. The final Gemini flight, Gemini 12 (11 November 1966), rendezvoused with its target Agena on the third orbit and kept station with it.

Would astronauts be able to perform useful work outside their spacecraft when in orbit or on the Moon? This was the question extravehicular activity (EVA) was designed to answer. The answers proved to be various and more difficult than had been envisioned.

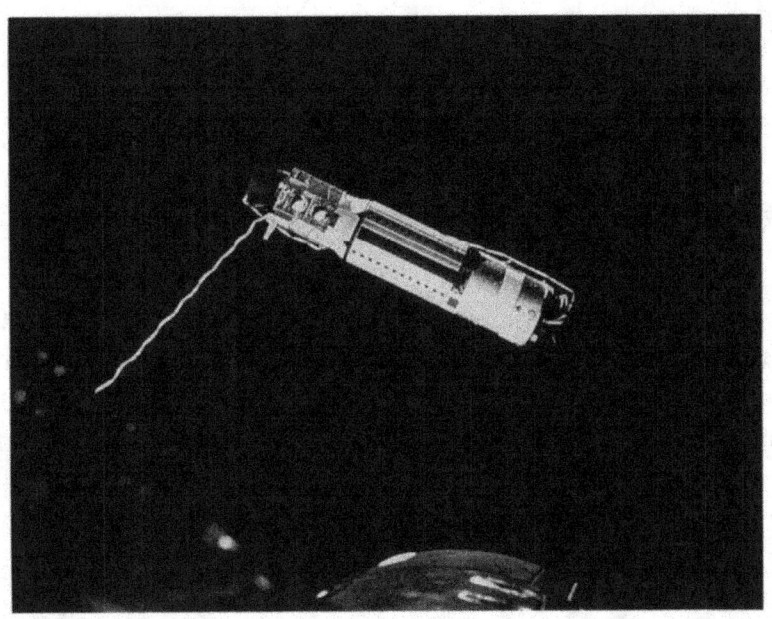

The view from Gemini 11's window of the Agena rocket with which the Gemini crew is practicing rendezvous and tethered station keeping.

Gemini 4 began EVA when Edward H. White II floated outside his spacecraft for 23 minutes. Protected by his spacesuit and attached to Gemini by a 26 foot umbilical cord, White used a handheld maneuvering unit to move about, took photographs, and in general had such an exhilarating experience that he had to be ordered back into the spacecraft. Because he had no specific work tasks to perform, his EVA seemed deceptively easy.

America's first space walk. Astronaut Edward H. White II fired short birsts with his hand-held maneuvering gun to move around in the zero gravity of space before returning to the Gemini 4 spacecraft.

That illusion was rudely shattered by the experience of Gemini 9, when Eugene A. Cernan spent 2 hours in EVA; he had tasks to perform in several areas on the spacecraft. His major assignment was to go behind the spacecraft into the adapter area, put on the 165-pound astronaut maneuvering unit --a more powerful individual flight propulsion system the Air Force had built-- and try it out. The effort to get the unit harnessed to his back was so intense that excessive perspiration within his spacesuit overtaxed the system and fogged his visor. The experiment was abandoned and he was ordered back into the spacecraft.

Much more pleasant was the experience of Michael Collins on Gemini 10. He tried two kinds of EVA: the first time he stood in the open hatch for 45 minutes and made visual observations and took pictures; the second time he went out on a 33-foot long tether, maneuvering for 55 minutes with the handheld maneuvering unit and even propelled himself over to the station-keeping Agena and removed a micrometeoroid-impact experiment which had been in space for four months. But reality raised its ugly head again during Gemini 11 when Richard F. Gordon, Jr., was assigned a full schedule of work tasks along the spacecraft but had to terminate after 33 minutes because of fatigue. He had battled himself to exhaustion trying to control his bodily movements and fight against the opposite torque that any simple motion set in train. It was Isaac Newton's Third Law of Motion in pure form.

NASA had learned its lesson. When Gemini 12 went up, many additional body restraints and hand and footholds had been added. Astronauts had trained for the strange floating sensation by doing the same assignments in water tanks on Earth. Results were gratifying; in a 2 hour 6 minute tethered EVA (aside from two standup EVAs) Edwin E. Aldrin, Jr., successfully performed 19 separate tasks. Total EVA on this flight added up to 5 hours 28 minutes.

On the last seven flights, Gemini experimented with the aerodynamic lift of the spacecraft to ensure pinpoint landings on Earth's surface; with the dispersions possible when Apollo came in from 230,000 miles away, tired astronauts would need this. The inertial guidance system provided inputs to the computer, which solved the guidance equations. On flights 6-10 the reentry was controlled by the crew. On the last two flights the data were fed into the automatic system. Results were promising. The average navigational accuracy of the seven flights was within 2 miles of the aiming point, much better than previous flights.

Gemini was primarily a technological learning experience. So it is not surprising that of the 52 experiments in the program, more than half (27) were technological, exploring the limits of the equipment. But there were also 17 scientific experiments and 8 medical ones. An important one was the 1400 color photographs taken of Earth from various altitudes. This provided the investigators the first large corpus of color photographs from which to learn more about the planet on which we live.

Probably the most valuable management payoff from Gemini was the operational one: how to live and maneuver in space; next was how to handle a variety of situations in space by exploiting the versatility and depth of the vast NASA-contractor team that stood by during flights. Finally there were valuable fiscal lessons: an advanced technology program had a "best path" between too slow and too fast. Deviation on either side, as had occurred in the early days of Gemini, could cost appalling amounts of money. But once on track, even economies were possible. Once Gemini flights were on track, for example, associate administrator for Manned Space Flight George E. Mueller (successor to Holmes) had won agreement from his principal contractors to cut the three-month period between launches to two months. This was primarily to get Gemini out of the way before Apollo launches started, but it paid off financially, too; where total program costs for Gemini were estimated in 1964 to be $1.35 billion, the actual cost closed out at $1.29 billion.

This, then, was Gemini, a versatile, flexible spacecraft system that wound up exploring many more nooks and crannies of spaceflight than its originators ever foresaw--which is as it should be. Major lessons were transmitted to Apollo; rendezvous, yes; docking, yes; EVA, yes; manned flights up to two weeks in duration, yes. Equally important, there was now a big experience factor for the astronauts and for the people on the ground, in the control room, around the tracking network, in industry. The system had proved itself in the pit; it had evolved a total team that had solved realtime problems in space with men's lives at stake. This was no mean legacy to Apollo.

Some of the technological payoff had come too late. With the increasing sophistication of Gemini and the consequent slippage of both financial and engineering schedules, the Apollo designers and engineers sometimes had to invent their own wheel. But the state of the art had been advanced: thrusters, fuel cells, environmental control systems, space navigation, spacesuits, and other equipment. In the development stage of Apollo the bank of knowledge from Gemini paid off in hundreds of subtle ways. The bridge had been built.

Boosters and Spacecraft for Apollo

Throughout Gemini's operational period, Apollo was slogging along toward completed stages and completed spacecraft. Saturn I, the booster almost overtaken by events, finished its 10-flight program in 1964 and 1965 with six launches featuring a liquid-hydrogen second stage. Not only was it proved out; the clustered-engine concept was demonstrated and an early form of Apollo guidance was tested. The last four flights were considered operational; one (18 September 1964) tested a boilerplate Apollo spacecraft. The last three carried Pegasus meteoroid-detection satellites into orbit. The last two Saturn I boosters were fabricated entirely by industry, making a transition from the Army-arsenal in-house concept that had previously characterized the Marshall Space Flight Center. Ten launches, ten successes.

Meanwhile the larger brother, the Saturn IB, was being born. Its first stage was to generate 1.6 million pounds of thrust, from eight of the H-1 engines that had powered Atlas and Saturn I, but uprated to 200,000 pounds each. The second stage was to feature the new J-2 liquid hydrogen engine, generating 200,000 pounds of thrust. It was a crucial element of the forthcoming Saturn V vehicle, since in a five-engine cluster it would power the second stage and a single J-2 would power the third stage.

Saturn IB was the first launch vehicle to be affected by a new concept, "all-up" testing. Associate Administrator Mueller, pressed by budgetary constraints and relying on his industry experience in the Air Force's Minuteman ballistic missile program, pressed NASA to abandon its stage-by-stage testing. With intensive ground testing of components, he argued, NASA could with reasonable confidence test the entire stack of stages in flight from the beginning, at great savings to budget and schedule. Marshall engineers had built their splendid success record by being conservative; they vigorously opposed the new concept. But eventually Mueller triumphed. On 26 February 1966, the complete Saturn IB flew with the Apollo command and service module in suborbital flight; the payload was recovered in good condition. On 5 July the IB second stage, the instrument unit-- which would house the electronic and guidance brains of the Saturn V--and the nose cone were propelled into orbit. The total payload was 62,000 pounds, the heaviest the U.S. had yet orbited. On 26 August a suborbital launch qualified the Apollo command module for manned flight; the attached service module fired its engine four times; and an accelerated reentry trajectory tested the Apollo heatshield at the 25,000-MPH velocity of a spacecraft returning from lunar distance.

The largest brother, Saturn V, was still being pieced together. Developed by three different contractors, the three stages of Saturn V had individual histories and problems. The first stage, although the largest, had a long lead-time and was on schedule. The third stage, though enlarged and sophisticated from the version flown on Saturn IB, had a previous history. It was the second stage that was the newest beast--five J2 engines burning liquid hydrogen. It became the pacing item of the Saturn V and would remain so almost until the first launch.

As manned space launches became more frequent, logistics became a major problem. Oversized cargoes like the Apollo instrument unit segment, as well as command modules and upper stages were carried by the Super Guppy, a dramatically modified Boeing Stratoliner.

Of the three spacecraft, the lunar module was, early and late, the problem child. For one thing, it was begun late--a whole year late. For another, it differed radically from previous spacecraft. There were two discrete spacecraft within the lunar module; one would descend to the lunar surface from lunar orbit; the other would separate from the descent stage and leap off the lunar surface into lunar orbit and rendezvous with the command module. The engine for each stage would have to work perfectly for that one time it fired. Both had teething troubles. The descent engine was particularly troublesome, to the point that a second contract was let for a backup engine of different design. Weight was a never-ending problem with the lunar module. Each small change in a system, each substitution of one material for another, had to be considered as much in terms of pounds added or saved as in any gain in system efficiency. By the end of 1966, the Saturn 1B and the block 1 Apollo command and service module were considered man-rated.

On 27 January 1967, AS-204, to be the first manned spaceflight, was on the launch pad at Cape Kennedy, moving through preflight tests. Astronauts Virgil I. Grissom, Edward H. White II, and Roger B. Chaffee were suited up in the command module, moving through the countdown toward a simulated launch. At T minus 10 minutes tragedy struck without warning. As Major General Samuel C. Phillips, Apollo program director, described it the next day: "The facts briefly are: at 6:31 p.m. (EST) the observers heard a report which originated from one of the crewmen that there was a fire aboard the spacecraft . . . ." Ground crew members saw a flash fire break through the spacecraft shell and envelop the spacecraft in smoke, Phillips said. Rescue attempts failed. It took a tortuous five minutes to get the hatch open from the outside. Long before that the three astronauts were dead from asphyxiation. It was the first fatal accident in the American spaceflight program.

Shock swept across the nation and the world. In the White House, President Johnson had just presided over the signing of an international space law treaty when Administrator Webb phoned with the crushing news. Webb said the next day: "We've always known that something like this would happen sooner or later....who would have thought the first tragedy would be on the ground?"

Who, indeed? What had happened? How had it happened? Could it happen again? Was someone at fault? If so, who? There were many questions, few answers. The day following the fire, Deputy Administrator Seamans appointed an eight-member review board to investigate the accident. As chairman he chose Floyd L. Thompson, the veteran director of the Langley Research Center. For months the board probed the evidence, heard witnesses, studied documentation. On 10 April, Webb, Seamans, Mueller, and Thompson briefed the House space committee on the findings: the fire had apparently been started by an electrical short circuit which ignited the oxygen- rich atmosphere and fed on combustible materials in the spacecraft. The precise wire at fault could probably never be determined. Like most accidents it should not have happened. There had been errors in design, faults in testing procedures. But the basic spacecraft design was sound. A thorough review of spacecraft design, wiring, combustible materials, test procedures, and a dozen more items was underway. Congress was not satisfied. Hearings in both houses continued, gradually eroding Webb's support on Capitol Hill.

The block I spacecraft would not be used for any manned flights. The hatch on the block II spacecraft would be redesigned for quick opening. The hundreds of miles of wiring in the spacecraft were checked for fire-proofing, protecting against damage, and other problems. An intensive materials research program devised substitute materials for combustible ones. In effect, the block II spacecraft was completely redesigned and rebuilt. The cost: 18 months delay in the manned flight schedule and at least $50 million. The gain: a sounder, safer spacecraft.

Well before men flew in Apollo spacecraft the question had been raised as to what, if anything, NASA proposed to do with men in space after Apollo was over. With the long lead-times and heavy costs inherent in manned space programs, advance planning was essential. President Johnson proposed the question to Webb in a letter on 30 January 1964. NASA's first-look answer surfaced in congressional hearings on the fiscal 1965 budget. Funds were requested for study contracts that would investigate a variety of ideas for doing new things in space with the expensively acquired Apollo hardware. Possibilities: long-duration Earth-orbital operations, lunar surface exploration operating out of an unmanned Apollo lunar module landed on the Moon, long-duration lunar orbital missions to survey and map the Moon, Earth-orbital operations leading to space stations.

Through 1965 and 1966 the studies intensified and options were fleshed out. The Woods Hole conference in the summer of 1965 brought together a broad spectrum of the American science community and identified some 150 scientific experiments that were candidates for such missions. By 1966 there was a sense of urgency in NASA planning; the Apollo production line was peaking and would begin to decline in a year or two. Unless firm requirements for additional boosters, spacecraft, and other systems could be delineated and funded soon, the production lines would shut down and the hard-won Apollo skills dispersed. In the fiscal 1967 congressional hearings, NASA presented further details and fixed the next fiscal year as the latest that hardware commitments could be deferred if the Apollo production line was to be used.

NASA went into the fiscal 1968 budget cycle with a fairly ambitious Apollo Applications proposal. It asked for an appropriation of $626 million as the down payment on six Saturn 1Bs, six Saturn Vs, and eight Apollo spacecraft per year. The Bureau of the Budget approved a budget request of $454 million. This cut the program by one-third. Congress appropriated only $253 million, so by mid-1968 the plan was down to only two additional Saturn 1Bs and one orbital workshop, with it and its Apollo telescope mount being deferred to 1971.

Spacecraft for Space Science

Manned spaceflight, with its overwhelming priority, had had both direct and indirect impact on the NASA space science program. From 1958 to 1963, scientific satellites had made impressive discoveries: the van Allen radiation belts, Earth's magnetosphere, the existence of the solar wind. Much of the space science effort in the next four years had been directed toward finding more detailed data on these extensive phenomena. The radiation belts were found to be indeed plural, with definite, if shifting, altitudes. The magnetosphere was found to have an elongated tail reaching out beyond the Moon and through which the Moon periodically passes. The solar wind was shown to vary greatly in intensity with solar activity.

All of these were momentous discoveries about our nearby space environment. The first wave of discoveries said one thing to NASA: if you put up bigger, more sophisticated, more versatile satellites than those of the first generation, you will find many other unsuspected phenomena that might help unravel the history of the solar system, the universe, and the cosmic mystery of how it all works. So a second generation of spacecraft was planned and developed; they were called observatory class --five to ten times as heavy as early satellites, built around a standard bus instrumented for a specific scientific discipline, but designed to support up to 20 discrete experimental instruments that could be varied from one flight to the next-- solar observatories, astronomical observatories, geophysical observatories. As these complex spacecraft were developed and launched in the mid-1960s, the first results were on the whole disappointing. The promise was confirmed by fleeting results, but their very complexity inflicted them with short lifetimes and electrical failures. There were solid expectations that these could be worked out for subsequent launches. But by the late 1960s the impingement of manned spaceflight budgets on space science budgets reduced or eliminated many of these promising starts. Smaller satellites, such as the Pioneer series, survived and made valuable observations, measuring the solar wind, solar plasma tongues, and the interplanetary magnetic field.

Lunar programs faired somewhat better but did not come away unscathed. The lunar missions were now in support of Apollo, so they were allowed to run their course. Surveyor softlanded six out of its seven spacecraft on the Moon from 1966 through 1968. Its television cameras gave Earthlings their first limited previews of ghostly lunar landscapes seen from the surface level. Its instruments showed that lunar soil was the consistency of wet sand, firm enough to support lunar landings by the lunar module. Lunar Orbiter put mapping cameras in orbit around the Moon in all of its five missions, photographed over 90 percent of the lunar surface, including the invisible back side, and surveyed potential Apollo landing sites.

Planetary programs suffered heavy cuts. The Mariner series was cut back, but its two flights provided exciting new glimpses into the history of the solar system. Mariner 4 flew past Mars on 14 July 1965 and gave us our first closeup view of Earth's fabled neighbor. At first glance the view was disappointing. Mars was battered by meteor impacts almost as much as the Moon. While there were no magnetic fields or radiation belts, there was a thin atmosphere. Mariner 5 flew past Venus on 19 October 1967; this second pass at mysterious Venus found no magnetic field but an ionosphere that deflected the solar wind. The atmosphere was dense and very hot; temperatures were recorded as high as 700 K, with 80 percent of the atmosphere being carbon dioxide. But the immediate future of more sophisticated planetary exploration seemed bleak. The ambitious Voyager program was curtailed in 1966 and finally dropped in 1968; it envisioned large planetary spacecraft launched on Saturn V which would deploy Mars entry capsules weighing up to 7000 pounds.

The applications satellites had been a crowning achievement for NASA in the early 1960s. The NASA policy of bringing a satellite system along through the research and development stages to flight demonstration of the system and then turning it over to someone else to convert into an operational system received its acid test in 1962. With the demonstration of Syncom performance, the commercial potential of communications satellites became obvious and immediate. NASA's R&D role seemed over, but how should the valuable potential be transferred to private ownership without favoritism? The Kennedy administration's answer was the Communications Satellite Corporation, a unique government-industry-international combination. The board of directors would be made up of six named by the communications industry, six by public stockholders, and three named by the President of the United States. The corporation would be empowered to invite other nations to share the investment, the services, and the profits. This precedent-setting proposal stirred strong political emotions, especially in the Senate. A 20-day debate ensued, including a filibuster, the time-honored last resort in cases of deeply divisive issues, before the administration proposal was approved. On 31 August 1962, President Kennedy signed the bill into law. ComSat Corp, as it came to be called, set up in business. On 6 April 1965, its first satellite, Early Bird I was launched into synchronous orbit by NASA on a reimbursable basis. By the end of 1968, there was an Intelsat network of five communications satellites in synchronous orbits, some 20 of an expected 40 ground stations in operation, and 48 member nations participating. The Soviets had mounted a competitive system of Molniya satellites with first launch in 1965. They too had sought international partnership, but only France outside of the Iron Curtain countries signed up. By 1968 they had launched 10 Molniya satellites into their standard elliptical orbit. On the American side, the question of government-sponsored research on communications satellites was not completely solved by the creation of ComSat Corp. Congress continued to worry over the thorny question of whether the government should carry on advanced research on communications satellites versus the prospect that a government-sponsored monopoly would profit from the results.

Weather satellites were simpler in the sense that the relationship was confined to two government agencies. The highly successful Tiros was seized on by the Weather Bureau as the model for its operational satellite series. NASA had high hopes for its follow-on Nimbus satellite, bigger, with more instruments measuring more parameters. The Weather Bureau, however, felt that unless NASA could guarantee a long operational lifetime for Nimbus, it was too expensive for routine use. So NASA continued Nimbus as a test bed for advanced sensors that could provide better measurements of the vertical structure of the atmosphere and global collection of weather data.

Navigational satellites, one of the early bright possibilities of space, continued to be intractable. But there was a new entry, the Earth resources satellite. Impressed by the Tiros photographs and even more by the Gemini photographs, the Department of Interior suggested an Earth resources satellite program in 1966. Early NASA investigation envisioned a small, low-altitude satellite in Sun-synchronous orbit. What could be effectively measured with existing sensors, to what degree, with what frequency, in what priority? These questions involved an increasing number of government agencies. Then there was the complex question of what trade-off was best between aircraft-borne sensors and satellite-borne ones. It was a new kind of program for NASA, involving many more government agencies and many more political sensitivities than the uncluttered researches in space.

Aspects of Flight Research

The advanced research activities of NASA also became more subtle and difficult to track. An interlocking network of basic and applied research, advanced research was designed to feed new ideas and options into the planning process. The most visible portion was flight research, which sometimes supported work in the space program.

Although ballistic reentry from space had become familiar by the 1960s, there was a group of engineers who argued in favor of "lifting" reentry. The idea was to build a spacecraft with aerodynamic characteristics so that a crew could fly back through the Earth's atmosphere and land at an airfield. The X-20A Dyna-Soar proposed by the Air Force was one such example.

But the Dyna-Soar never flew, a victim of budget constraints and new technology. The NACA became involved in a smaller series of lifting body aircraft that helped pave the way for the Space Shuttle design. At Ames, a series of exploratory studies during the 1950s culminated in a design known as the M2, a modified half-cone (it was flat on the top) and a rounded nose to reduce heating. NASA engineers at Edwards kept up with much of the theoretical ideas percolating out of Ames, and Robert Reed became fascinated by the M2, by now called the "Cadillac" for the two small fins emerging at the blunt tail. He built a successful flying model, which led to authorization for a manned glider.

In many ways, the local authorization was more typical of the early NACA, since Headquarters did not know about it --nor did Langley, for that matter. But it seemed promising and it could be done cheaply. One aircraft company later estimated it would have cost at least $150,000 to build the M2, but the Edwards crew did it for less than $50,000. A nearby sailplane company built the laminated wooden shell (Reed was also an avid sailplane pilot); a considerable amount of other fabrication work was done by NASA personnel who were practiced hobbyists in the art of homebuilt aircraft. The landing gear was scrounged from a Cessna 150. By 1963, the M2-F1, as it was now called, had been completed.

Initial flight tests required a ground vehicle to tow the M2-F1 above the dry lake bed, but none of NASA's trucks or vans was fast enough for the task. The Edwards team had to shop around for a hopped-up Pontiac convertible, further modified by a custom car shop in Long Beach to include rollbars, radio equipment, and special seats for observers. Results from the ground tow tests were good, so the next step involved aerial tow tests behind a C-47. By the time these flights concluded in 1964, the lifting-body concept, despite its oddball history, seemed to be worth pursuing. NASA Headquarters and congressional people were both impressed. News reporters loved the lifting-body saga, and there was keen interest in the more advanced lifting-body designs already under consideration.

Three lifting body configurations grouped on the dry lake bed at Flight Research Center. Left to right: the X-24, M-2, and HL-10.

The M2-F1 showed the way, but far more work was needed, involving high-speed descent and landing approach tests. By this time, the Air Force was interested, and a joint lifting-body program was formalized in 1965. Generally speaking, NASA, through the Flight Research Center at Edwards, held responsibility for design, contracting, and instrumentation, while the Air Force supplied the launch aircraft for drop tests, assorted support aircraft, medical personnel, and the rocket power plant to be used in the advanced designs.

Northrop became the prime contractor for the aluminum "Heavyweights" sponsored by NASA. The M2-F2 was a similar, but refined version of the M2-F1; Northrop also delivered the HL-10, which had a very short, angled delta wing and a different fuselage shape. There was progress as well as disappointment; a landing accident destroyed the M2-F2 and cost the pilot the sight of one eye. The plane was rebuilt as the M2-F3 with an additional vertical fin. The HL-10,with a flat bottom and rounded top fuselage became the most successful, capable of Mach 1.86 speeds and altitudes of 90,000 feet. At a time when arguments over a "deadstick" shuttle reentry became hottest, some crucial HL-10 landing tests convinced planners that a shuttle without special landing engines could successfully complete reentry, approach, and landing. A final confirmation came during tests of the Martin X-24A (based on an Air Force project), whose shape was similar to a laundry iron. By the time that the X-24A test flights ended (1969-71), designers had complete confidence in the ability of the space shuttle to land on a conventional runway at the end of a space mission. The lifting-body tests made an important contribution.

In other projects, explicit aeronautical research continued. At the Flight Research Center, another exotic plane captured the attention of flight aficionados --the Rockwell XB-70 Valkyrie, a Mach 3 high-altitude bomber. The Air Force began plans for the XB-70 in 1955, but by the time of its rollout ceremonies in 1964, plans for a fleet of such large bombers had given way to reliance on advanced ICBMs with more powerful warheads. In the meantime, the Kennedy administration had endorsed studies for a supersonic transport (SST) for airline use, and the configuration of the XB-70 made it an excellent candidate for flight tests in support of the SST program.

The North American XB-70 yielded valuable data on flight characteristics of large, supersonic aircraft. This photo of a test flight shows shock waves and vortices forming on the fuselage and wings.

The XB-70 Valkyrie took to the air for the first time in the autumn of 1964. With a fuselage length of 189 feet and a large delta wing measuring 105 feet from tip to tip, its size, operating characteristics, and construction features made it an excellent SST prototype. The Air Force and NASA began a cooperative test program with the XB-70 in the spring of 1966, the first airline-sized aircraft in the world able to make sustained, long-range supersonic flights. The flight requirements for a Mach 3 airliner similar to the XB-70 were far more complicated than those for a Mach 2 aircraft, such as the Anglo-French Concorde SST. A Mach 3 airliner's structure required more exotic alloys, such as titanium, because the conventional aluminum airframe of a plane like the Concorde could not survive the aerodynamic heating at greater speeds. Integrating a Mach 3 aircraft into the existing airway traffic system became a special problem, because it made turns that required hundreds of miles to complete. Working with the XB-70 uncovered a number of operational and maintenance problems.

Despite the loss of one XB-70 in a midair collision, killing two test pilots, the NASA test program generated invaluable data on sustained supersonic flight. On one hand, XB-70 tests conclusively demonstrated that shock waves from SST airliners would prohibit supersonic routes over the continental United States. These tests helped fuel the opposition to the American SST program. On the other hand, the knowledge accumulated about handing qualities and structural dynamics represented basic data for use in future supersonic military aircraft and in high-speed airliners. But the test program was too expensive to sustain indefinitely. Early in 1969, the XB-70 Valkyrie made its last flight, to the Air Force Museum in Dayton, Ohio.

When the political question arose as to whether the United States should enter the international competition for a supersonic commercial transport aircraft --a sweepstakes already begun by Great Britain and France jointly with their Concorde and by the Soviet Union with its TU-144-- NASA already had a solid data base to contribute. It also had the laboratories and the contracting base to manage the program. But wise counsel from Deputy Administrator Dryden led to NASA's retreat into a supportive R&D role; he argued that with Apollo underway, NASA could not politically sponsor another high-technology, enormously expensive program during the same budget years without one of them being sacrificed to the other or killing each other off in competition for funds. The subsequent history of the SST program, including its eventual demise, was eloquent testimonial to the wisdom of his judgment. His death in December 1965 was a loss to the nation's aerospace program.

Other research efforts paid big dividends within the space program. Lewis Research Center had become involved in the use of liquid hydrogen as a rocket fuel in 1955. Although liquid hydrogen offered very attractive increases in thrust per pound as compared to previous fuels, hydrogen had a bad reputation left over from dirigible days and the Hindenburg disaster. But by 1957 Lewis was successfully and routinely firing a 20,000-pound thrust engine using liquid hydrogen as fuel. It was these tests that gave NASA the confidence in 1959 to decide that the upper stages of the lunar rocket should be fueled with liquid hydrogen. Without this additional rocket power, it might have been impossible (or at least much more expensive) to put men on the Moon.

Long-range prospects of manned planetary exploration depended heavily on more efficient thrust per pound of fuel propulsion. To this end NASA had continued the long-range program inherited from the Air Force to develop a nuclear-propelled upper stage for a rocket. Engineering down to a compact package the enormous weight, size, and shielding of the kind of reactor used in nuclear electric power plants was a severe challenge. The inevitable intensification of radiation density and temperatures defeated existing materials that would contain and transmit the heat to an engine. Time after time over the years, test firings of promising configurations had to be stopped prematurely when radiation corrosion took its toll. Finally in December 1967 the NRX-A6 reactor ran for one hour at full power, twice the time achieved before. Improvements in reactor fuel elements cut radiation control in half. The SNAP program of radioisotope thermoelectric generators also progressed. The SNAP-27 was the longlife power source for the Apollo science experiments to be left on the lunar surface.

Apollo to the Moon

Although the tragic fire of January 1967 delayed plans for manned spaceflight in Apollo hardware for approximately 18 months, the versatility of the system came to the rescue. The burden of checking out the major components of the system was quickly shifted to unmanned flights while a quick-opening hatch was designed and tested, combustibles were sought out and replaced, and the wiring design was completely reworked. After a nine-month delay, flight tests resumed. On 9 November 1967, Apollo 4 became the first unmanned launch of the awesome Saturn V. A 160-foot high stack of three-stage launch vehicle and spacecraft, weighing 2824 tons, slowly lifted off Launch Complex 39, propelled by a first- stage thrust of 7.5 million pounds. A record 278,000 pounds of payload and upper stage were put into Earth orbit. Later the third stage fired to simulate lunar trajectory, lifting the spacecraft combination to over 10,000 miles. With the third stage discarded, the service module fired its engine to raise the apogee to 11,000 miles, then burned again to propel the spacecraft toward Earth reentry at the 25,000 MPH return speed from the Moon. All systems performed well; the third stage could restart in the vacuum of space; the automated Launch Complex 39 functioned beautifully. The once-controversial concept of "all-up" testing had been vindicated.

Next came the unmanned flight test of the laggard lunar module. On 22 January 1968, a Saturn 1B launched a 32,000-pound lunar module into Earth orbit. It separated, and tested its ascent and descent engines. The lunar module passed its first flight test.

Now to man-rate the huge Saturn V. Apollo 6, on 4 April 1968, put the launch vehicle through its paces--the stages, the guidance system, the electrical systems. Four of five test objectives were met; Saturn V was man-rated. The scene was set for the first manned spaceflight in Apollo since the tragic fire. Apollo 7 would test the crew and command module for the 10 days in space that would later be needed to fly to the Moon, land, and return.

But beyond Apollo 7, the schedule was in real difficulty. It was the summer of 1968; only a year and a half remained of the decade within which this nation had committed itself to land astronauts on the Moon. Somehow the flight schedule ought to be accelerated. Gemini's answer had been to launch missions closer together, but the size and complexity of Apollo hardware severely limited that option. The only other possibility was to get more done on each flight. For a time, however, it seemed that the next flight, Apollo 8, would accomplish even less than had been planned. It had been scheduled as the first manned test of the lunar module in Earth orbit, but the lunar module had a lengthy test-and-fix roadblock ahead of it and could not be ready before the end of the year, and perhaps not then. So a repeat of Apollo 7 was considered, another test of the command module in Earth orbit without the tardy lunar module but this time on the giant Saturn V. Eight years earlier that would have been considered a big bite; now, was it big enough, given Apollo's gargantuan task?

In Houston, George Low didn't think it was. After all, he reasoned, even this test-flight hardware was built to go to the Moon; why not use it that way? The advantages of early experience at lunar distances would be enormous. On 9 August he broached the idea to Gilruth, who was enthusiastic. Within days the senior managers of the program had been polled and had checked for problems that might inhibit a circumlunar flight. All problems proved to be fixable, assuming the Apollo 7 went well. The trick then became to build enough flexibility into the Apollo 8 mission so that it could go either way, Earth-orbital or lunar-orbital.

Apollo 7 was launched on 11 October 1968. A Saturn IB put three astronauts into Earth orbit, where they stayed for 11 days, testing particularly the command module environmental system, fuel cells, communications. All came through with flying colors. On 12 November, NASA announced that Apollo 8 had been reconfigured to focus on lunar orbit. It was a bold jump.

On 21 December a Saturn V lifted the manned Apollo 8 off Launch Complex 39 at the Cape. The familiar phases were repeated: Earth orbit, circularizing the orbit, all as rehearsed. But then the Saturn third stage fired again and added the speed necessary for the spacecraft to escape Earth's gravity on a trajectory to the Moon. All the rehearsed or simulated steps went well. On 23 December the three-man crew became the first human beings to pass out of Earth's gravitational control and into that of another body in the solar system. No longer were humans shackled to the near environs of Earth. The TV camera looked back at a small, round, rapidly receding ball, warmly laced with a mix of blue oceans, brown continents, and white clouds that was startling against the blackness of space.

As Apollo 8 came around the backside of the Moon after going into lunar orbit, the crew was greeted with this haunting view of the earth rising above the desolate lunar horizon.

On Christmas Eve Apollo 8 disappeared behind the Moon and out of radio communication with Earth. Not only were the astronauts the first humans to see the mysterious back side of the Moon; while there they had to fire the service module engine to reduce their speed enough to be captured into lunar orbit --irrevocably, unless the engine would restart later and boost them back toward Earth.

Another engine burn regularized their lunar orbit at 70 miles above the surface. Television shared the breathtaking bird's eye view of the battered lunar landscape with hundreds of millions on Earth. The crew members read the creation story from Genesis and wished viewers a Merry Christmas. On Christmas Day they fired the service module engine once again, acquired the 3280 feet per second additional speed needed to escape lunar gravity, and triumphantly headed back to Earth. They had at close range verified the lunar landing sites as feasible and proved out the hardware and communications at lunar distance, except for the all-important last link, the lunar module.

That last link, the lunar module, was still of major concern to NASA. Two more flights were expended to confirm its readiness for lunar landing. The Apollo 9 flight (3-13 March 1969) was the first manned test of the lunar module. The big Saturn V boosted the spacecraft combination into Earth orbit. The lunar flight drill was carefully rehearsed; the command and service modules separated from the third stage of the Saturn V, turned around, and docked with the lunar module. The lunar module fired up and moved away to 113 miles; then the spacecraft rendezvoused and docked.

A final test --was anything different at lunar distance? On 18 May 1969, Apollo 10 took off on a Saturn V to find out. The entire lunar landing combination blasted out to lunar distance. Once in lunar orbit, the crew separated the lunar module from the command module, descended to within 9 miles of the surface, fired the ascent system, and docked with the command module. Now all systems were "go."

Astronaut Neil A. Armstrong took this photograph of Edwin E. Aldrin, Jr., deploying the passive seismic experiments at tranquility base, while the ungainly lunar module crouches in the background.

On 16 July 1969, Apollo 11 lifted off for the ultimate mission of Apollo. Saturn V performed beautifully. The spacecraft combination got off to the Moon. Once in lunar orbit, the crew checked out their precarious second home, the lunar module. On 20 July the lunar module separated and descended to the lunar surface. At 4:18 P.M. (EST) came the word from Astronaut Neil A. Armstrong: "Houston,--Tranquility Base here. --The Eagle has landed." After checkout, Armstrong set foot on the lunar surface: "one small step for a man --one giant leap for mankind." The eight- year national commitment had been fulfilled; humans were on the Moon. Armstrong set up the TV camera and watched his fellow astronaut Edwin E. Aldrin, Jr., join him on the lunar surface, as Michael Collins circled the Moon in the Columbia command module overhead. More than one-fifth of the Earth's population watched ghostly TV pictures of two spacesuited men plodding around gingerly in an unlikely world of gray surface, boulders, and rounded hills in the background. The astronauts implanted the U.S. flag, deployed the scientific experiments to be left on the Moon, collected their rock samples, and clambered back into the lunar module. The next day they blasted off in the ascent module and rendezvoused with the command module.

The astronauts returned to an ecstatic reception. For a brief moment, people's day-to-day divisions had been suspended; the world watched and took joint pride in this achievement in exploration. Astronauts and their families made a triumphant world tour which restated world pride in this new plateau of humanity's conquest of the cosmos.

Chapter 6

AEROSPACE DIVIDENDS (1969-1973)

The worldwide euphoria over mankind's greatest voyage of exploration did not rescue the NASA budget. At its moment of greatest triumph, the space program was being drastically cut back from the $5 billion budgets that had characterized the mid-1960s. Part of the reduction was expected; the peak of Apollo production line expenses was past. But the depth of the cut stemmed from emotional changes in the political climate, mostly centering on the unpopular Vietnam war--its sapping expenses in lives and money, the debilitating protests at home. As Congress read the public pulse, the cosmos could wait; the Soviet threat had for the moment been put to rest; the new political reality lay in domestic problems. NASA's fiscal 1970 budget was reduced to $3.7 billion. Something had to give. The basic Apollo mission was continued, but the last three flights had to be deleted. Space science projections were hit hard. The ambitious $2 billion Voyager program for planetary exploration dwindled into oblivion; it would later resurface as the much more modest Viking. The new Electronics Research Center in Cambridge, Massachusetts, under construction since 1964, was transferred to the Department of Transportation intact--a $40 million facility taking with it 399 of 745 skilled employees.

Space Probes and Earth Satellites

But the bought and paid for projects continued to earn dividends. An Orbiting Astronomical Observatory (OAO-2) was launched 7 December 1968. It was the heaviest and most complex automated spacecraft yet in the space science program. It took the first ultraviolet photographs of the stars. The results were portentous: first hard evidence of the existence of "black holes" in space. Mariner 6 and Mariner 7, launched in early 1969, journeyed to Mars, flew past as close as 1900 miles, took 198 high-quality TV photos of the planet, 2000 ultraviolet spectra, and 400 infrared spectra of the atmosphere and surface.

Other programs continued with prepaid momentum. The fifth and sixth Orbiting Solar Observatories (OSO) were launched in 1969, as was the sixth Orbiting Geophysical Observatory. In 1970 Uhuru was launched and scanned 95 percent of the celestial sphere for sources of x-rays. It discovered three new pulsars in addition to the one previously identified. In 1971 Mariner 9 was launched; on 10 November, the first American spacecraft went into orbit around another planet. The early months in orbit were discouraging; a gigantic dust storm covered most of the martian surface for two months. But the dust gradually cleared; photographs in 1972 showed startling detail. Mapping 85 percent of the martian surface, Mariner 9 photographs depicted higher mountains and deeper valleys than any on Earth. The rocky martian moons, Deimos and Phobos, were also photographed. OSO-7, launched on 29 September 1971, was the first satellite to catch on film the beginning of a solar flare and the consequent streamers of hot gases that extended out 10.6 million kilometers; it would also discover "polar ice caps" on the sun (dark areas thought to be several million degrees cooler than the normal surface temperatures). With the confirmation of black holes, the enigmatic collapsed star remnants so dense in mass and gravity that even light cannot escape, and the previous discoveries of quasars and pulsars, these findings added up to the most exciting decade in modern astronomy.

Planetary exploration opened further vistas of other worlds. Pioneer 10, launched 2 March 1972, left the vicinity of Earth at the highest velocity ever achieved by a spacecraft (32,000 MPH) and took off on an epic voyage to the huge, misty planet Jupiter. Giant of the solar system, swathed with clouds, encircled by a cluster of moons, Jupiter was an inescapable target if one hoped to understand the composition of the solar system. Out from the Sun, out from Earth, Pioneer 10 ventured for a year and a half, through the unexplored asteroid belt and far beyond. After a 992 million kilometer journey, on 3 December 1973 the tiny spacecraft flew past Jupiter. It survived the fierce magnetic field and sent back photographs of the huge planet and several of its moons, measured temperatures and radiation and the magnetic field. Steadily sailing past Jupiter and away from the Sun, in 1987 Pioneer 10 would cross the orbit of Pluto, becoming the first man-made object to travel out of our solar system and into the limitless reaches of interstellar space.

Pioneer 10's partner, Pioneer 11, took off on 5 April 1973 to follow the same outward path. On 3 December 1974 it passed Jupiter at the perilously close distance of 26,000 miles--as opposed to 80,000 for Pioneer 10--and returned data. The composite picture from the reports of the two spacecraft depicted an enormous ball of hydrogen, with no fixed surface, emitting much more radiation than it received from the Sun, shrouded with a turbulent atmosphere in which massive storms such as the Great Red Spot (25,000 miles in length) had raged for at least the 400 years since Galileo first trained a telescope at Jupiter. Pioneer 11 swung around the planet and, taking advantage of Jupiter's gravitational field, accelerated outward at 66,000 MPH toward the distant planet Saturn, where in 1979 it would observe at close range this lightest of the planets (it could float on water), its mysterious rings, and its 3000 mile diameter moon Titan.

Going in the other direction, Mariner 10 left Earth on 3 November 1973, headed inward toward the Sun. In February 1974 it passed Venus, gathering information that confirmed the inhospitable character of that planet. Then using Venus's gravitational force as propulsion, it charged on toward the innermost planet, Mercury. On 29 March 1974, Mariner 10 flew past Mercury, providing man a 5000 times closer look at this desolate, crater-pocked, sun-seared planet than had been possible from Earth. Using the gravitational field of its host planet to alter course, Mariner 10 flew out in a large elliptical orbit, circled back by Mercury a second time on 21 September 1974, and a third time on 16 March 1975. The cumulative evidence pictured a planet essentially unchanged since its creation some 4.5 billion years ago, except for heavy bombardment by meteors, with an iron core similar to Earth's, a thin atmosphere composed mostly of helium, and a weak magnetic field.

Fascinating as the information about our fellow voyagers in the solar system was and as important as the long-range scientific consequences might be, Congress and many government agencies were much more intrigued with the tangible, immediate-return, Earth-oriented program that began operations in 1972. On 23 July ERTS 1 (Earth Resources Technology Satellite) was launched into polar orbit. From that orbit it would cover three-quarters of the Earth's land surface every 18 days, at the same time of day (and therefore with the same sun angle for photography), affording virtually global real-time information on developing events such as crop inventory and health, water storage, air and water pollution, forest fires and diseases, and recent urban population changes. In addition it depicted the broad area (and therefore undetectable by ground survey or aircraft reconnaissance) geologic patterns and coastal and oceanic movements. ERTS 1 also interrogated hundreds of ground sensors monitoring air and water pollution, water temperature and currents, snow depth, etc., and relayed information to central collection centers in near real-time. The response was instantaneous and wide-spread. Foreign governments, states, local governments, universities, and a broad range of industrial concerns quickly became involved in both the exploration of techniques to exploit these new wide-area information sources and in real-time use of the data for pressing governmental and industrial needs. Some 300 national and international research teams pored over the imagery. For the first time accurate estimates were possible of the total planting and growth status of wheat, barley, corn, and rice crops at various times during the growing season; real-time maps versus ones based on data that would have been collected over a period of years; timber cutting patterns; accurate prediction of snow runoff for water management; accurate, real-time flood damage reports. Mid-term data included indications that the encroachments of the Sahara Desert in Africa could be reversed by controlled grazing on the sparse vegetation in the fringe areas; longer range returns suggested promise in monitoring strip mining and subsequent reclamation, and in identification of previously unknown extensions of Earth faults and fractures important to detection of potential earthquake zones and of associated mineral deposits.

Like the experimental communications satellites of the early 1960s, the ERTS found an immediate clientele of governmental and commercial customers clamoring for a continuing inflow of data. The pressure made itself felt in Congress; on 22 January 1975, Landsat 2 (formerly ERTS 2) was orbited ahead of schedule to ensure continuation of the data that ERTS 1 (renamed Landsat 1) had provided for two and a half years, and a third satellite was programmed for launch in 1977. This would give confidence to experimental users of the new system that they could securely plan for continued information from the satellite system.

The Earth resources program had another important meaning. It was a visible sign that the nature and objectives of the space program were undergoing a quiet but dramatic shift. Where the Moon had been the big target during the 1960s and large and expensive programs had been the name of the game, it became increasingly clear to NASA management as the decade ended that the political climate would no longer support that kind of a space program. The key question now was, "What will this project contribute to solving everyday problems of the person in the street?" One by one the 1960s-type daydreams of big, away-from-Earth projects were reluctantly put aside: a manned lunar base, a manned landing on Mars, an unmanned "grand tour" of several of the planets. When the Space Shuttle finally won approval, it was because of its heavy dedication to studies of our Earth and its convincing economies in operation.

Another sign of the times was that NASA was increasingly becoming a service agency. In 1970 NASA for the first time launched more satellites for others (ComSat Corp, NOAA, DoD, foreign governments) than for itself. Five years before only 2 of 24 launches had been for others. Clearly this trend would continue for some years.

Twilight for Apollo

Meanwhile Apollo was running its impressive course. Apollo 12 (1424 November 1969) repeated the Apollo 11 adventure at another site on the Moon, the Ocean of Storms. One attraction of that site was that Surveyor3 had been squatting there for two and a half years. A pinpoint landing put the lunar module within 600 feet of the Surveyor spacecraft. In addition to deploying scientific instruments and collecting rock samples from the immediate surroundings, Astronauts Conrad and Bean cut off pieces from Surveyor 3, including the TV camera, for return to Earth and analysis after 30 months of exposure to the lunar environment.

Apollo 13 was launched 11 April 1970, to continue lunar exploration. But 56 hours into the flight, well on the way to the Moon, there was a "thump" in the service module behind the astronauts. An oxygen tank had ruptured. Pressure dropped alarmingly. What was the total damage? Had other systems been affected? How crippled was the spacecraft combination? The backup analysis system on Earth sprung into action. Using the meager data available, crews at contractor plants all over the country simulated, calculated, and reported. The verdict: Apollo 13 was seriously, perhaps mortally, wounded. There was not air or water or electricity to sustain three men on the shortest possible return path to Earth. But, ground crews and astronauts asked simultaneously, what about the lunar module, a self-contained spacecraft unaffected by the disaster? The lunar landing was out of the question anyway; the lifesaving question was how to get three men around the Moon and back to Earth before their life-supporting consumables ran out. Could the lunar module substitute for the command module, supplying propulsion and oxygen and water for an austere return trip? The simulations said yes. Apollo 13 was reprogrammed to loop around the Moon and set an emergency course for Earth return. The descent engine for the lunar module responded nobly; off they went back to Earth. It was a near thing --powered down to the point of minimum heating and communication, limiting activity to the least possible to save oxygen. Again the flexibility and depth of the system came to the rescue; when reentry was safely within the limited capabilities of the crippled Apollo, the "lifeboat" lunar module was jettisoned along with the wounded service module.Apollo 13 reentered safely.

The next flight was delayed while the causes and fixes for the near-tragedy on Apollo 13 were sorted out. On 31 January 1971, Apollo 14 lifted off, the beginning of the scientific exploration of the Moon. The major new system was a transporter, a cart on which to load equipment and bring back rock samples. A major target of the Apollo 14 mission to Fra Mauro was to climb the walls of the Cone Crater; the attempt was halted as time ran out and the astronauts had trouble pinpointing the location.

Apollo 15 introduced the Moon car, the lunar rover. With this electric-powered, four-wheel drive vehicle developed at Marshall at a cost of $60 million, the astronauts roamed beyond the narrow confines of their landing site and explored the area. Astronauts on this flight covered 17 miles of lunar surface, visited a number of craters in the Hadley-Apennines area, and photographed the ghostly ravine Hadley Rille. Thanks to the lowered exertion level because of the lunar rover, exploration time was doubled.

The remaining Apollo missions now had all the equipment planned for lunar exploration. Apollo 16 landed in the Descartes area in April 1972, stayed 71 hours, provided photos and measurements of lunar properties. Apollo 17, launched 7 December 1972, ended the Apollo program with the most productive scientific mission of the lunar exploration program. The site, Taurus-Littrow, had been selected on the basis of previous flights. Objectives were to seek out both oldest and youngest rocks to fill in the geologic history of the Moon. For the first time a trained geologist, Harrison H. Schmitt, was on a crew, adding his professional observations. EVA time was over 22 hours and the lunar rover traveled some 22 miles.

Apollo was ended. From beginning to end, it had lasted 11 1/2 years, cost $23.5 billion, landed 12 men on the Moon, and produced an unassessable amount of evidence and knowledge. Technologically it had produced hardware systems several orders of magnitude more capable than their predecessors. In various combinations, the components of this technology could be used for a wider variety of explorations than the nation could possibly afford. The luxury of choice was, which of a half-dozen possible missions?

Scientific answers were going to be returned over several decades. The Lunar Receiving Laboratory had been constructed in Houston to be the "archive" of the 840 pounds of physical lunar samples that had been returned from various parts of the Moon by six lunar landing crews. Scientists in this country and 54 foreign countries were analyzing the samples with an impressive variety of instruments and the expertise of many scientific disciplines. Gross results had already established that the Moon was a separate entity from Earth, formed at the same time as Earth some 4.5 billion years ago; that it had its own volcanic history; that with no protective atmosphere it had been bombarded for eons by meteors from outer space, which had plowed up the surface lava flows from the lunar interior. Refinement of data would go on for decades.

Apollo had proved many other things: the ability of a diversified system of government, industry, and universities to mobilize behind a common national purpose and produce on schedule an immense and diverse system directed to a common purpose. It not only argued that society could do many things in space, whether extended lunar exploration from permanent lunar bases or manned excursions to Mars, but argued that solutions to many of humanity's major problems on Earth --pollution, food supply, and natural disasters such as earthquakes and hurricanes could be ameliorated or controlled by the combination of space technology and the large-scale management techniques applied to it.

Next in manned spaceflight came Skylab. Trimmed back to one orbital workshop and three astronaut flights, Skylab had had a hectic financial and planning career, the converse of Apollo. The revised plan called for an S-IVB stage of the Saturn V to be outfitted as two-story orbiting laboratory, one floor being living quarters and the other working room. The major objective of Skylab was to determine whether humans could physically withstand extended stays in space and continue to do useful work. Medical data from the Gemini and Apollo flights had not completely answered the question. Since there would be far more room in the 89 foot long orbital workshop than in any previous spacecraft, William C. Schneider, Skylab program director, devised a more extensive experiment schedule than all previous spaceflights combined. Most ambitious in terms of hardware was the Apollo telescope mount; five major experiments would cover the entire range of solar physics and make it the most powerful astronomical observatory ever put in orbit. The other major areas of experimentation were Earth resources observations and medical experiments involving the three-man crew. There were important subcategories of experiments: the electric furnace, for example, would explore possibilities of using the weightless environment to perform industrial processes that were impossible or less effective on 1g Earth, such as forming perfectly round ball bearings or growing larger crystals, much in demand in the electronics industry.

On 14 May 1973 a giant Saturn V lifted off from Kennedy Space Center to place the unmanned 165,000 pound orbital workshop in Earth orbit. Within minutes after launch, disquieting news filtered through the telemetry reports from the Saturn V. The large, delicate meteoroid shield on the outside of the workshop had apparently been torn off by the vibrations of launch. In tearing off it had caused serious damage to the two wings of solar cells that were to supply most of the electric power to the workshop; one of them had sheared off, the other was snagged in the folded position. Once the workshop was in orbit, the news worsened. The loss of the big shade exposed the metal skin of the workshop to the hot sunshine; internal temperatures soared to 325 K. This heat not only threatened its habitation by astronauts, but if prolonged might fog sensitive film and generate poisonous gases.

The launch of the first crew was twice postponed, while the far-flung ground support team worked around the clock for 10 frantic days, trying to improvise fixes that would salvage the $2.6 billion program. With only partial knowledge of the precise degree and nature of the damage, engineers had to work out fixes that met the known problems, yet were versatile enough to cope with unknown ones. There were two major efforts: first, to devise a deployable shade that the astronauts could spread over the metal surface of the workshop; the other was to devise a versatile tool kit of cutters and snippers to release the solar wing from whatever prevented it from unfolding.

On 25 May 1973, an Apollo command and service module combination was lifted into orbit by a Saturn 1B. Apollo docked with the workshop on the 25th. The crew entered it the next day and deployed a makeshift parasol through the solar airlock. The effect was immediate; internal temperature began to drop. On 7 June Astronauts Conrad and Kerwin clambered outside the workshop and after a tense struggle succeeded in cutting the metal straps that ensnared the remaining solar wing; it slowly deployed and electrical power poured into the storage batteries. Human ingenuity and courage had made the workshop operational again.

The remaining Skylab missions were almost anticlimactic after the dramatic rescue of the workshop. With only minor problems, the missions ticked off their complicated schedules of experiments. In spite of the initial diversion, the first crew obtained 80 percent of the solar data planned; 12 of 15 Earth resources runs were completed; and all of the 16 medical experiments went as planned. Its 28-day mission completed, the crew undocked and returned to Earth.

The second crew was launched on 28 July 1973, completed almost 60 days in orbit, and exceeded by one-third the solar observations and Earth resources runs planned. All the medical experiments were performed. The third crew (launched 16 November 1973) completed an 84-day flight with all experiments performed, as well as the additional observations of the surprise cosmic visitor, comet Kohoutek.

The vast mass of astronomical and Earth resources data from the Skylab program would take years to analyze. A more immediate result was apparent in the medical data and the industrial experiments. With the corrective exercises available on Skylab, there seemed to be no physiological barrier to the length of time humans could survive and function in space. Biological functions did indeed stabilize after several weeks in zero-g. The industrial experiments gave strong evidence that the melting and solidification process was promisingly different in weightlessness; single crystals grew five times as large as those producible on Earth. Some high-cost industrial processes apparently had new potential in space.

As the empty Skylab continued to circle the Earth, its orbit began to decay, threatening an uncontrolled reentry. NASA regained some control over the rogue Skylab in the spring of 1979, and managed to steer it to reentry over the Indian Ocean. Still, chunks of the Skylab made a fiery plunge into remote areas of Australia, a reminder of the potential dangers of civilization's own debris from space.

Transonic and Hypersonic Flight Research

Although questions about an SST aircraft persisted, NASA and its principal contractor, Boeing, kept working on the design throughout the 1960s. By 1971, production plans were under way when the program came to a halt. Critics remained adamant about the costs of the SST and its ability to operate economically. Flight tests of the big XB-70 Valkyrie had done little to quell the issue of sonic booms, and there were worrisome questions about adverse environmental effects at high altitudes. Congress finally voted against funds for construction of an SST for flight testing.

The British and French proceeded with a smaller SST, the jointly developed Concorde, which began flight tests in 1969 and entered service in 1976. A Soviet SST, the Tupolev TU-144, also began internal schedules in 1976, but was withdrawn from service two years later. Meanwhile, NASA and American aerospace companies cooperated in a research effort known as the Supersonic Cruise Aircraft Research Program. Beginning in 1973, this activity involved analysis of propulsion systems and advanced airframes. Continuing into the 1980s, the ongoing SST studies made considerable progress in quieter, cleaner engines as well as much improved passenger capacity and operational efficiencies. If the opportunity for second-generation SST airliners materialized later, NASA and the aerospace industry intended to lead the way with an American design.

While investigation of the supersonic regime continued, a major breakthrough at the transonic level occurred --the supercritical wing. The transonic regime had beguiled aerodynamicists for years. At transonic speeds, both subsonic and supersonic flow patterns encased an aircraft. As the flow patterns went supersonic, shock waves flitted across the wings, resulting in a sharp rise in drag. With most commercial jet airliners operating in the transonic range, coping with this drag factor could bring major improvements in cruise performance and yield substantial benefits in operating costs.

During the 1960s, Richard Whitcomb committed himself to a program intended to resolve the transonic problem. For several years, Whitcomb intensely analyzed what came to be called the "supercritical" Mach number--the point where the airflow over the wing went supersonic, with a resultant decline in drag. Analysis and wind tunnel tests led to a wing with a flattened top surface (to reduce its tendency to generate shock waves) and a downward curve at the trailing edge (to help restore lift lost from the flattened top). But wind tunnel tests were one thing. Real planes in the air were often something else. The next step meant thorough flight testing of a plane equipped with the unusual wing.

Fortunately, NASA came up with an available plane that lent itself to comparatively easy modification: the Vought F-8A Crusader. The structure of the plane's shoulder-mounted wing made it easy to remove and replace with the supercritical design. Moreover, the F-8A was built with landing gear that retracted into the fuselage, leaving the experimental wing with no outstanding production encumbrances. The Navy had spare planes available, and its speed of Mach 1.7 made it ideal for transonic flight tests. Although the test plane had begun life as a Navy fighter, the supercritical wing program was aimed at civil applications. The airlines as well as the airline manufacturers closely followed development of the new airfoil.

The modified Crusader, designated the TF-8A, made its first flight at Edwards in 1971 and continued for the next two years. The test flights yielded data that corresponded to measurements from the preliminary tunnel tests at Langley. Most important, the supercritical wing promised genuine improvement in the transonic region, a fact that translated directly into reduced fuel costs and lower operational costs. Ironically, foreign manufacturers of business jets were the first to apply the new technology in new designs like the Canadair Challenger (Canada) and the Dassault Falcon (France). At the same time, both Boeing and Douglas applied the concept in experimental Air Force transports like the YC-14 and YC-15.

As additional commercial manufacturers began utilizing data from the supercritical wing studies, NASA and the Air Force collaborated in the development of its military applications for combat planes. Known as TACT, for Transonic Aircraft Technology, the military effort used a modified F-111A. By the early 1980s, with refined flight testing of the F-111A still continuing, several operational aircraft had been designed to utilize information from this project.

NASA's use of military aircraft to probe the transonic region paralleled a different effort that involved very high supersonic speeds. The aircraft in this case was one of the most exotic creations to fly --the Lockheed YF-12A, a highly classified interceptor design that led to the equally highly classified SR-71A Blackbird reconnaissance aircraft. According to published performance figures, the Blackbirds were capable of Mach 3 speeds at altitudes of 80,000 feet or more. The planes originated in the famed Lockheed "Skunk Works" of Clarence "Kelly" Johnson, where Johnson and a talented group of about 200 engineers put aeronautical pipe dreams on paper, and then proceeded to build and fly them. The operating requirements of the plane at extreme speeds and altitudes for sustained periods created a completely new regime of requirements for parts and systems. As Johnson commented later, "everything on the aircraft from rivets and fluids, up through the materials and power plants, had to be invented from scratch."

The first Blackbird flew in 1962; NASA first became involved in 1967, when Ames, where early wind tunnel data was acquired under tight security, was given permission to use the data in ongoing research. In return the Flight Research Center at Edwards organized a small team to assist the Air Force flight tests. But NASA wanted its own Blackbird for tests that would support the SST program still under way in the late 1960s. By this time, the SR-71A was operational, and the Air Force had put two YF-12A prototypes in storage at Edwards. When the Air Force offered the pair to NASA, the agency quickly accepted and also assumed operational expenses as well, although the Air Force assigned a small team for assistance in maintenance and logistics.

NASA launched its Blackbird program with great enthusiasm. Engineers from Lewis, Langley, and Ames had a keen interest in propulsion research, aerodynamics, structural design, and the accuracy of wind tunnel predictions involving Mach 3 aircraft. The first YF-12A test missions under NASA jurisdiction began late in 1969 and flights averaged once a week during the next 10 years, examining an impressive variety of high-speed problems. One series involved a biomedical team who monitored physiological changes in the flight crews in order to measure stress in the demanding environment of high-speed operations. Many Blackbird test flights routinely carried instruments to analyze boundary layer flow, skin friction, heat transfer, and pressures in flight. Various structural techniques were employed in test panels on the planes. An experimental computerized checkout system diagnosed problems in flight and provided information for required maintenance prior to the next mission. The checkout system was seen as a valuable one for application in the Space Shuttle as well as military and commercial planes.

In many ways, the Blackbird program, covering a decade of intensive flight tests, was one of the Flight Research Center's most useful programs, with a rich legacy of information for later aircraft built for sustained cruise at Mach 3. The end of the program prompted a chorus of protest from the Blackbird flight team and other NASA personnel who felt the United States was frittering away its lead in high-speed flight and in technology generally. Such grumbling was probably premature. The interest in aerospace and a national commitment to new technology was still high, although it took different directions. At first glance, the new concern for controlling aircraft noise, reducing pollutants from engines, and enhancing overall aircraft fuel efficiency might have seemed less glamorous than derring-do at Mach 3. But the rationale for confronting such issues became urgent in the late 1970s, and the solutions to these issues were no less complex and challenging than the problems of high-speed flight. Aeronautical research continued to be a dynamic field of NASA programs to come.

Chapter 7

ON THE EVE OF SHUTTLE (1973-1980)

While Skylab was being built, other events significant to the future of space exploration were taking place. The initiatives bore the imprint of Thomas O. Paine, acting administrator after Webb's resignation in 1968 and administrator of NASA from March 1969 until he returned to industry in September 1970. One goal was a broad approach to increased cooperation in space exploration. As had so many of our international space initiatives in the postwar period, this effort offered separate proposals to the Soviet Union and to Western European countries. The approach to the Soviet Union began in 1968, with suggestions for advanced cooperation, especially in the expensive arena of manned spaceflight. One area of Soviet vulnerability might be rescue of astronauts and cosmonauts. By now the Soviet Union had lost four cosmonauts in flight, three in one accident, one in another. They had always evidenced a singular concern for cosmonaut safety. Perhaps some joint program could develop a system of international space rescue. The dynamics seemed right; by 1969 the evidence was clear that, whether the Soviet Union had in fact been in a moon-landing race with the United States, the United States was ahead. Secrecy in space was virtually nonexistent; size of payloads, destinations of missions, performance --all were detectable by tracking systems.

Paine's first offer was for Soviet linkup with the Skylab orbital workshop. But the very hardware implied inequity. The Soviets were not interested. Further explorations found lively Soviet interest in a completely new project to develop compatible docking and rescue systems for manned spaceflight. Negotiations proceeded rapidly. Completed by George M. Low, acting administrator after Paine's departure, the grand plan for the Apollo-Soyuz Test project (ASTP) called for a mutual docking and crew exchange mission that could develop the necessary equipment for international rescue and establish such criteria for future manned systems from both nations. A Soyuz spacecraft would lift off from the Soviet Union and establish itself in orbit. Then an Apollo spacecraft would be launched to rendezvous and dock with the Soviet craft. Using a specially developed docking unit between the two spacecraft, they would adjust pressurization differences of the two spacecraft and spend two days docked together, exchanging crews and conducting experiments. All of this was agreed to and rapidly became a significant test for the validity of the detente agreements which President Richard M. Nixon had negotiated with the Soviet Union.

An unprecedented detailed cooperation between the two superpowers ensued. A series of joint working groups of Soviet and American specialists met over several years to work out the various hardware details and operational procedures. At the Nixon-Brezhnev summit in 1973, the prospective launch date was narrowed to July 1975. The most concrete example of U.S.-U.S.S.R. cooperation in space proceeded with good faith on both sides. The mission flew as scheduled on 15 July and smoothly fulfilled all objectives.

The Space Shuttle

The other major initiative of Paine's began on the domestic front and then expanded to the international arena. Skylab having been narrowed to the point that it would be a limited answer to the future of manned spaceflight, President Nixon appointed a Space Task Group to recommend broad outlines for the next 10 years of space exploration. Within this group, Paine won acceptance for the concept of the Space Shuttle. In its original conception, the Space Shuttle would have been a rocket-boosted airplane-like structure that would take off from a regular airport runway, fly to orbital speed and altitude, deploy satellites into orbit, repair or retrieve satellites already in orbit, and, using an additional Space Tug stage, lift manned and unmanned payloads throughout the solar system. Compared to earlier methods, the big changes would be that the launcher and Shuttle would be reusable for up to 100 flights, halving the cost per pound in orbit. But subsidiary changes were only slightly less important: satellites could be designed for orbital rigors, not the additional ones of rocket launch. In a manned mission, the Shuttle would handle a crew of up to seven people in orbit; three of these could be non-pilot scientists who went along to operate their experiments in an unpressurized laboratory carried in the Shuttle cargo bay. The flight crew alone could deliver 65,500 pounds of assorted satellites into orbit.

The Space Task Group submitted its report to the President on 15 September 1969. It offered three levels of effort: option 1 would feature a lunar-orbital station, an Earth-orbital station, and a lunar surface base in the 1980s; option 2 envisioned a Mars manned mission in 1986; option 3 included initial development of space station and reusable shuttles but would defer landing on Mars until some time before the end of the century. Eventual peak expenditures on these options were estimated to vary from $10 billion down to $5 billion per year. Study and rework went on for more than two years. Paine left NASA to return to industry; his successor, James C. Fletcher, took office in April 1971 and immediately reviewed the status of the Space Shuttle, particularly for its political salability. He became quickly convinced that the Shuttle as then envisioned was too costly to win approval. Total costs for its development were estimated at $10.5 billion. Fletcher instigated a rigorous restudy and redesign which cut the cost in half, mainly by dropping the plan for unassisted takeoff and substituting two external, recoverable, reusable solid rockets and an expendable external fuel tank. This proved to be salable; President Nixon approved the development of the Space Shuttle on 5 January 1972.

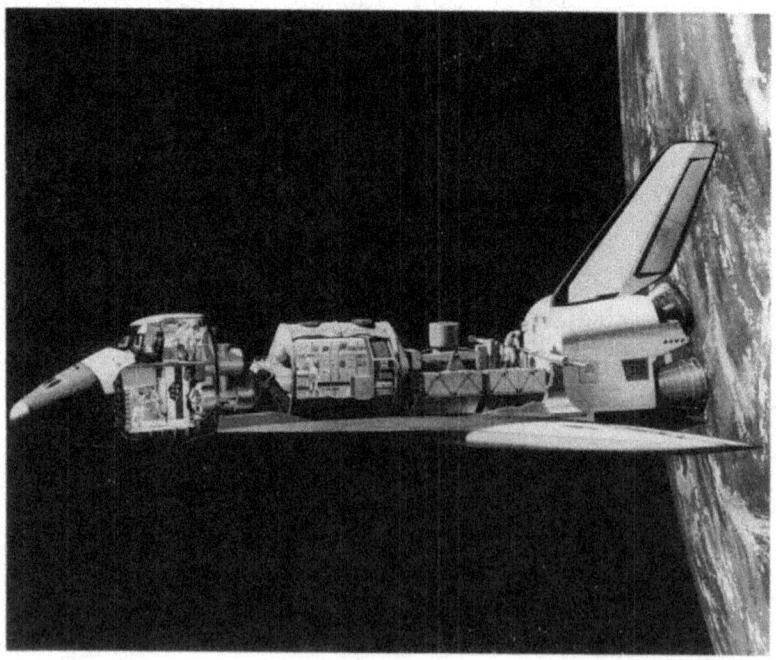

In this cutaway illustration, the Shuttle orbiter is shown with the ESA Spacelab as the prime payload. Scientific instruments are mounted on ESA-built pallets mounted in the rear of the Shuttle's cargo bay.

First Paine and then Fletcher had been trying to get a commitment from Western European nations for a major system in the Shuttle. Their own joint space program had not been an unqualified success. In 1964, Western European nations had joined to form two international space organizations, ELDO to produce launch vehicles and ESRO to produce spacecraft and collect and interpret results. The technical capability was there, but issues of assigning specific contracts to separate countries and allocating budgets hampered rapid European progress. A proposed booster had three stages, each developed in a different country. The launch record was a gloomy history of one kind of failure after another. After years of effort, Western Europe had little to show for its independent launch vehicle. On the other hand, much had been learned about multinational coordination of advanced technology, and successful joint projects like Concorde and several multinational military aircraft ventures (such as the Pavavia Tornado) had promoted a sophisticated aerospace community in Europe. Moreover, using American boosters, the ESRO group had successfully launched a variety of scientific satellites, applications satellites, and space probes. In addition to experienced contractors, the European space organizations had developed international centers, like ESTEC in the Netherlands, to carry out research and maintain ongoing management of space projects. By the early 1970s, there was general agreement on the need for a new, unified organization. Based on the strengthening capabilities of its aerospace community, the European Space Agency (ESA) was established in 1975. A new start was in the air.

It was into this restive environment that Paine came to talk about the next generation of the U.S. space program and to hold out promise of some discrete major segment to be developed and produced in Europe --a partnership that would give them a meaningful piece of the action with full pride of useful participation. Europe's response was warm, though it took a while to coalesce. Finally the joint decision was made: Western Europe agreed to build the self-contained Spacelab that would fit in the cargo bay of the Shuttle spacecraft; a pressurized module would provide a shirtsleeve environment for scientists to operate large-scale experiments; an unpressurized scientific instrument pallet would give large telescopes and other instruments direct access to the space environment. The cost, an estimated $370 million. In 1975 Canada joined the international effort, agreeing to foot the $30 million R&D bill for the remote manipulator used to emplace and retrieve satellites in orbit.

The Space Shuttle promised a whole new way of spaceflight: nonpilots in space; multiple payloads that could be placed where they were wanted or picked up out of orbit; new designs of satellites, free from the expensive safeguards against the vibrations and shocks of launch by rocket. The $5.2 billion program would buy two prototypes for test in 1978 and 1979. Projected flight programs from 1980 to 1991 identified a total of almost 1000 payloads to be handled by the Shuttle.

The largest consumer of the NASA budget and of management attention during the late 1970s was the Space Shuttle. Since its beginnings in the early 1970s, the development story for the Space Shuttle had been quite different from that of Apollo in the 1960s. The original projected costs had been halved to win the necessary political approval of the program; this cut was only achieved by making severe compromises in the original design --from a system that would take off from a runway like an airplane, fly into orbit, and return to land on a runway like an airplane, to a system that would take off vertically like a rocket, jettison the boosters and fuel tanks, and return to land on a runway like an airplane. This initial compromise was not to be the last, as the budget continued to be lean year after year. Potential development problems were worked around because the money was not available to investigate them. The consequences of this insufficient level of research during the development cycle were not apparent in the years when the Shuttle was being designed and the components fabricated. As late as 1977, when the orbiter Enterprise was carried aloft by a modified Boeing 747 and dropped to make approach and landing flights at Dryden Flight Research Center, progress was seen to be sure, if a little slow.

The Shuttle orbiter's descent/landing tests were launched from the Boeing 747, also used to carry orbiters between Kennedy, Edwards Air Force Base, and other sites.

In 1978 it became obvious that serious problems were dogging the main engines. A cluster of three of these high-pressure liquid-hydrogen-fueled engines would propel the orbiter into orbit, aided by two solid-rocket boosters. Not only were the main engines expected to produce the highest specific impulse of any rocket engine yet flown, but they also had to be throttleable and reusable --to fire again and again for many flights before being replaced. By 1979, a series of painstaking component-by-component analyses had identified and fixed most of the problems and individual engines were experiencing better test runs; but the first firings of the clustered engines generated a new set of problems. Grudgingly these too yielded to concentrated engineering rework; by the end of 1980 the total requirements of 80,000 seconds of test firing was in hand.

The other pacing item on the orbiter was the thermal protection tiling that would shield most of the orbiter surface from the searing heat of reentry. Manufacture and application of the 33,000 tiles lagged so badly that early in 1979 NASA decided to ferry the orbiter from the manufacturer's plant in California to Kennedy Space Center so that the remainder of the tiles could be applied there while other work and system checks were being done. But problems continued. The tiles were brittle and easily damaged; they did not bond to the metal properly and thousands had to be reapplied; they were too fragile and thousands more had to be removed, made more dense, and reapplied. Between the tiles and the engines, the Space Shuttle budget overran for several years and the date for the first flight slipped two painful years, with serious consequences for many government, domestic, and international customers. By the end of 1980, however, first flight in the spring of 1981 seemed truly possible. Operational flights were solidly booked out to the middle of the 1980s and the other three orbiters were moving through manufacturing.

Viking Orbiter montage of 102 photos of Mars in February 1980 (left) shows the Valles Marineris bisecting the planet, a gorge that would stretch from coast to coast of North America; to its left, three large volcanoes poke up through the unusual cloud cover.

The Planets

In space science the big program was Viking, which represented the first major fruit of a decision NASA had made some years before: to focus the space science program on the planets. Apollo, the reasoning went, would keep scientists busy for years analyzing the mass of data and samples that had been returned from the Moon. Not until that information had been assimilated would there be a need to consider whether more information was needed from the Moon and, if so, what kind.

Meanwhile space science, while not neglecting the study of the Sun and the universe, would concentrate on the inner planets of our solar system and begin an assault on the enigmatic outer planets. Apollo had shown, and the early planetary flights had confirmed, that every celestial body had worthwhile lessons to teach -- lessons that were important in their own right as science as well as lessons that illuminated problems on Earth. Why did Earth have the kinds and proportions of minerals that it had? Why tectonic plates and volcanism? Why oceans and the unique atmosphere of Earth? Why did our atmosphere circulate and transfer heat the way it did? Every new body we studied represented a new laboratory and a different set of data.

So it was that Mars, the most likely of the inner planets, became the first target of the more ambitious planetary program. In two launches the Viking program proposed to deploy four spacecraft in the vicinity of Mars; two orbiters would photograph the surface and serve as communications relays, while two landers would descend to the martian surface and photograph the terrain, measure and monitor the atmosphere and climate, and conduct chemical and biological tests on the soil for evidence of rudimentary life forms. It was very ambitious technology and complex science to be operated from over 40 million miles distance. But perform Viking did, in a technological triumph equal to (and in some ways greater than) the Apollo landings on the Moon. Arriving in the vicinity of Mars in mid-1976, the spacecraft went into orbit around the planet. Subsequently the two landers arced down to the rock-strewn surface where each landed safely. The two orbiters circled the planet, mapping most of the surface. That surface depicted by the orbiters, plus the weather and seismic reports from the landers, told a story of a planet with a quiescent present but a very different, active past. Volcanoes half again as high as any on Earth and great eroded canyons deeper and longer than any on Earth spoke of times, probably three billion years ago, when Mars was very active volcanically, with widespread liquid flows. Trace gases in the present thin atmosphere indicated a much denser atmosphere in the past. There was water, frozen in the polar ice caps; there were occasional dust storms; there were seasonal as well as diurnal variations in temperature; there was only a trace of seismic activity now. Viking's elaborate biology instruments detected no evidence of life forms. When the intensive one-year study of the planet ended, the spacecraft continued observations and reporting at intervals, providing further data on surface features, climate, and weather.

Earth's nearest planetary neighbor, Venus, was also probed during the last half of the 1970s. Two Pioneer spacecraft were launched toward Venus in the summer of 1978. Studying Venus presented a notably different problem than Mars or Earth. Its thick, heavy, hot atmosphere was impervious to normal photography and could be "seen" through only by means of radar. The first spacecraft arriving at Venus in December 1978, therefore, was an orbiter equipped with mapping radar to delineate the major features on the surface. The second spacecraft was a bus which released four probes in a broad pattern; these parachuted slowly through the atmosphere, sending back measurements until they crashed. The venusian atmosphere, they reported, was remarkably similar in composition and temperature on the day and night sides. There was a high sulfur content, with oxygen and water vapor at lower levels. By 1980 the orbiter had mapped over 80 percent of the venusian surface. Major features resembled two continents and a massive island chain--except there was no ocean. Instead a rolling plain enveloped the planet. One continent and the island chain were in the northern hemisphere. The continent was the size of Australia and had mountains taller than Everest; the island chain was apparently composed of two massive shield volcanoes more extensive than the Hawaii-Midway complex. The continent in the southern hemisphere was about half the size of Africa and exposed the lowest elevations on Venus in the Great Rift Valley, a huge trench 174 miles wide and 1395 long, with a depth similar to the great rift on Mars.

Study of the outer planets using more sophisticated spacecraft began in 1977 with the launch of Voyager 1 and 2 on 18-month flights to Jupiter. The Voyager system, Science magazine reported, was improved by a factor of 150,000 times over the Mariner 4 system, which flew to Mars in 1965. Voyager 1 made its closest approach to Jupiter in March 1979, with Voyager 2 following in July. The sensors recorded in fine-grain detail the intricate weather patterns on Jupiter and detected massive lightning bolts in the cloud tops. Passes by the Galilean moons revealed startling differences; active volcanoes on Io, ancient rings on Callisto marking the edges of huge impact craters. Europa's surface was laced with cracks from crustal movement, and Ganymede had a varying grooved and cratered surface.

With a boost from Jupiter's gravitational field, the Voyagers set course for distant, ringed Saturn, where Voyager 1 arrived in November 1980 and Voyager 2 arrived in August 1981. With sufficient control gas remaining, the mission extended to a faraway Uranus flyby in January 1986, with a Neptune flyby planned for August 1989. The venerable Pioneer II had visited Saturn in September 1979, discovering faint rings outside those discernible from Earth and demonstrating a safe flight path for Voyager 2 to follow on its path to Uranus.

In the study of the Sun and its interrelationships with Earth, NASA continued analysis of the mass of data acquired by Skylab's Apollo telescope mount. OSO 8, launched in 1975, to make a detailed study of the minimum phase of the 11-year solar cycle, returned data until 1978. Helios 2, part of a joint program with the Federal Republic of Germany to study the basic solar processes, was launched in 1976. As the solar cycle moved toward its maximum phase, the Solar Maximum Mission was launched in 1980 to study solar flares in the wavelengths in which the Sun releases most of its energy. Problems with the satellite led to rendezvous and retrieval by a Shuttle crew in 1984.

Voyager 1 and 2 photographs of Jupiter and its moon Io. Above, the violent weather patterns that constantly swirl around the edges of the Great Red Spot, the huge storm which is larger than Earth. Below, the vivid surface of Io, punctured with volcanoes and stained with their flow.

To study the effects of solar radiation on Earth's magnetosphere and atmosphere, NASA launched International Sun-Earth Explorer 1 and 2 in 1977. Positioned some distance apart but in similar elliptical orbits, the two satellites (one provided by NASA, the other by the ESA) monitored the complex interactions of Earth's magnetosphere with incoming solar radiation. In 1978 ISEE 3 was added to the system. Positioned much farther out from Earth, the spacecraft receives the solar wind and flares about an hour earlier, when they are unaffected by the magnetosphere.

In study of the universe, the major program of the second half of the 1970s was the series of three high-energy astronomy observatories. HEAO 1, launched in 1977 and the heaviest scientific satellite to date, surveyed the sky for x-ray sources, identifying several hundred new ones. HEAO 2, following the next year, studied in detail the most promising of those sources. HEAO 3, launched in 1979, surveyed the sky for gamma-ray sources and cosmic-ray flux. The other satellite orbited for study of the universe was the International Ultraviolet Explorer (IUE). Carrying instruments from NASA, the United Kingdom, and the ESA, IUE recorded ultraviolet emissions using two ground control centers from which the experimenters could direct the observations of the satellite much as is done with telescopes in observatories on Earth.

An intensified activity for NASA in the latter half of the 1970s was the congressionally mandated study of Earth's upper atmosphere, to learn more about the effects of gases such as freon on the ozone layer. A continuous measuring program resulted; several agencies provided data from which a detailed model of the complex processes could be constructed. The space applications program was active in the late 1970s. Communications research continued with the launch in 1976 of Communications Technology Satellite 1. A joint project with Canada, CTS 1, investigated the possibilities of high-powered satellites transmitting public service information to small, inexpensive antennae in remote locations.

Landsat 3 was launched in 1978, providing continuity for the flow of data to a growing number of users of Earth resources information. The most ambitious new Earth resources program was in agriculture. Encouraged by the results of the experimental Large Area Crop Inventory Experiment that ended in 1978 after demonstrating 90 percent accuracy in predicting the wheat production in the U.S. Southern Great Plains and U.S.S.R., the Department of Agriculture, with technical assistance from NASA and NOAA, began AgRISTARS (Agriculture and Resources Inventory Surveys Through Aerospace Remote Sensing).

A new form of resources surveying was attempted in 1978 with the launch of Seasat 1. Intended to report on such variables as sea temperature, wave heights, surface-wind speeds and direction, sea ice, and storms, Seasat 1 was an instant success. Unfortunately its life was cut short after three months in orbit by electrical power failure. Enough data had been recorded, however, to verify the effectiveness of the instrumentation and the existence of a group of potential users in the weather, maritime, and fisheries communities.

In environmental research, NASA launched Nimbus 7 in 1978, the last of the series of large experimental weather satellites. One of its instruments, together with one on Nimbus 4 and the observations of SAGE (Stratospheric Aerosol and Gas Experiment, launched in 1979), provided a profile and model of the ozone layer. The nations's weather satellite system was augmented in 1978 by the launch of Tiros-N and NOAA 6, the first two of a new generation of improved weather satellites in near-polar orbit. Tiros-N was a principal U.S. contributor to the international Global Atmospheric Research Program.

In geophysical research, a small experimental Heat Capacity Mapping Mission satellite was launched in 1978 to derive day and night temperatures of rock formations as a possible means of locating mineral-bearing strata. In 1979 another small satellite, Magsat, went into low orbit to take finer scale readings of anomalies in Earth's magnetic field that are directly related to crustal structure and therefore to possible mineral deposits. In earthquake research, NASA completed in 1979 the fourth phase of data gathering along the San Andreas Fault in California. By means of satellites ranging from specified points along both sides of the fault, experimenters estimated that the tectonic plates were moving 2.4 to 4.8 inches per year.

Aircraft and the Environment

In keeping with rising energy concerns of the 1970s, NASA committed considerable resources to new engine and aircraft technologies to increase flight efficiency as a means of conserving fuel. The Aircraft Energy Efficiency program was begun in 1975 to develop fuel-saving techniques that would be applicable to current aircraft as well as future designs. The project covered several areas of investigation: more efficient wings and propellers; composite materials that were lighter and more economical than metal; improved fuel efficiency in jet engines; new engine technologies for aircraft in the future.

The super critical wing was only one aspect of activity that also led NASA into the arcane subject of laminar flow-control. A smooth flow of air over the surface of a plane, or laminar flow, is a characteristic of low speeds. At cruising speeds, the air flow becomes turbulent, creating increased drag. Using models and analytical testing, NASA developed a system of tiny holes on the wing surface and a lightweight suction system to draw off the turbulent air. By the late 1980s, the agency was ready to begin flight testing of a laminar flow-control system for possible use on commercial aircraft.

Other research efforts were carried out through the Engine Component Improvement Program. The objective was to target engine components for which wear and deterioration led directly to decreased fuel efficiency in jet engines. As a result, new components to resist erosion and warping were introduced, along with improved seals, ceramic coatings to improve performance of gas-turbine blades, and improved compressor design. Research results were so positive and so rapidly adaptable that new airliners of the early 1980s like the Boeing 767 and McDonnell Douglas MD-80 series used engines that incorporated many such innovations.

For business jets, NASA rebuilt an experimental turbofan, incorporating newly engineered components designed to reduce noise. Completed by 1980, this project successfully developed engines that generated 50 to 60 percent less noise than current models. For larger transports, Lewis Research Center started tests of two research engines that cut noise levels by 60 to 75 percent and reduced emissions of carbon monoxide and unburned hydrocarbons as well.

In a different context, NASA became engaged in procedures for flight operations in increasingly congested air space. Among the issues that needed assessment were aircraft noise during landing and takeoff over populated areas; safe approach and landing procedures in bad weather; and methods for controlling high-density traffic patterns. Useful information emerged from a modified Boeing 737 twin-jet transport. In the plane's passenger area, NASA technicians put together a second cockpit equipped with the latest innovations in instrumentation. This second cockpit became the flight center for research operations; the crew occupying the standard cockpit in the 737's nose functioned as a backup. In addition to precision descent and approach procedures on instruments, the plane played a key role in demonstrating the Microwave Landing System in 1979. The International Civil Aviation Organization eventually adopted the Microwave Landing System over a competing European design to be used as the standard system around the world.

At Ames, scientists became interested in using aircraft as platforms for investigations of terrestrial as well as astronomical phenomena. Beginning in 1969, Ames acquired a number of different research planes and launched several imaginative investigations that continued over the following decades. High-altitude missions relied on a pair of Lockheed U-2 aircraft, originally supplied to the Air Force as reconnaissance planes. They carried out Earth resources observations, compiled land usage maps, surveyed insect infested crops, and measured damage from floods as well as forest fires. The highflying U-2 aircraft provided information covering hundreds of square miles; for a more intensive look at details in a smaller area Ames brought in other specialized planes that flew mid-altitude missions.

One of the pioneers in mid-altitude missions was a refurbished airliner--a Convair 990 christened the Galileo. Commencing operations in the early 1970s, the four-engine jet conducted a variety of tasks, such as infrared photography, detection of forest fires, and meteorological investigations. Over the Bering Sea in 1973, a joint study with the Soviet Union gathered data on meteorological phenomena, ice flow, and wildlife migratory patterns. The first Convair was lost in a tragic midair collision with a Navy patrol plane, but its operations had been so productive that acquisition of a second plane was authorized, and Galileo II went to work in 1974. Conducting research at mid-altitude heights, the new Convair 990 made international missions as well, including archaeological studies of Mayan ruins and observations of monsoon patterns in the Indian Ocean.

Other planes were added, like the small Learjet and the huge Lockheed C-141 Starlifter, which became operational with the Ames fleet in 1974. The Starlifter's interior size and load-carrying capacity made it the best candidate for installation of a 915-centimeter telescope for astronomical observations. Many of the C-141 missions, as well as those involving other Ames research planes, were international in scope. In 1977, the C-141, known as the Kuiper Airborne Observatory, flew to Australia to make observations of the planet Uranus during especially favorable astronomical conditions. American and Australian scientists studied the planet's atmosphere, composition, shape, and size, and discovered that Uranus possessed equatorial rings.

At about the same time, the Learjet, equipped with a 30-centimeter infrared telescope, was operating high over the Arctic on a different international mission. Known as Project Porcupine, Ames worked with the Max Planck Institut fur Physik und Astrophysik in a study of the coupling between the magnetosphere and the ionosphere. The experiment called for the launch of a sounding rocket from Sweden. After the rocket ejected a barium charge, the Learjet followed the barium trail along the Earth's magnetic lines of force. Collectively, these researches by aircraft on a global scale enhanced professional contacts for NASA personnel and generated favorable foreign press coverage for the agency as well as for the United States.

As Ames proceeded to carve out its niche in using aircraft as research platforms, the center also strengthened its role in flight research, moving beyond wind tunnel testing to flight testing. Taking advantage of Congressional support for aeronautical research, the director of Ames, Hans Mark (appointed 1969), guided the center into research on short-haul aircraft, including V/STOL designs. Since the mid-1960s, Ames had been working with the U.S. Army on helicopter research, relying on the big lowspeed tunnels at Ames, along with its excellent simulator equipment and other facilities. By the 1970s, both the FAA and the Air Force were working with Ames on a new generation of short-takeoff transports. In 1976, to the chagrin of Langley, Ames officially became NASA's lead center in helicopter research. Although the Pioneer project and future planetary missions shifted to the JPL at the same time (completed by 1980), the new aircraft programs enlivened activities at Ames.

Among the rotor craft investigations, one of the most interesting involved the XV-15 Tilt Rotor Research Aircraft, with wingtip-mounted engines. For takeoff and landings, the engines remained vertical, with the big rotors providing lift; once in the air the engines and rotors tilted to the horizontal, propelling the XV-15 forward. Bell Helicopter Textron built two aircraft for NASA and the Army. The first XV-15 went to Ames in 1978 for extensive tests in the 40 x 80 foot wind tunnel, to be followed by flight tests at Bell's plant in Texas. The first demonstration of inflight tests of the two prototypes was underway at Ames and at Dryden Flight Research Center during 1980.

The XV-15 tilt rotor research aircraft. For takeoff (above), the crafts rotors are horizontal to provide lift, then they pivot forward (bottom) to a full verticle position to give crusing speeds twice those of conventional helicopters.

Somewhat more conventional was the Quiet Short-Haul Research Aircraft to investigate new technologies for commercial airliners. The research plane was a hybrid, using an extensively modified de Havilland C-8A Buffalo. Under contract to NASA, Boeing rebuilt the plane with new avionics, new wings and tail, and a quartet of jet engines mounted above the wing to generate "upper surface blowing" in order to increase lift. The plane made its maiden flight at Boeing's Seattle plant in 1978, then flew to Ames for continued flight tests. The short takeoffs and quiet operations of the aircraft yielded much information for application in both civil and military design. One intriguing series of tests led to a successful landing and takeoff from an aircraft carrier-- the first four-engine jet plane to accomplish this feat.

For NASA, the decade of the 1980s seemed particularly promising. Its aeronautical programs were successful; space science had seen solid achievements; and progress in the Space Shuttle raised confidence for prospects of outstanding missions to come. That confidence was to be severely tested.

Chapter 8

AEROSPACE FLIGHTS (1980-1986)

For NASA flight research, the 1980s opened with a significant administrative change --the Dryden Flight Research Center lost its independent status and became a directorate of Ames Reseach Center in 1981. This did not mean that NASA was downgrading flight research; on the contrary, several exotic programs emerged during the decade, and a variety of unusual aircraft continued to populate the skies above Edwards.

Given the cost of experimental flight aircraft and the evolution of increasingly sophisticated electronic and simulator systems, it was perhaps inevitable that NASA eventually turned to smaller, pilotless radio-controlled aircraft. In the 1980s, this idea was embodied in the HiMAT, a contraction of Highly Maneuverable Aircraft Technology. The HiMAT, powered by a General Electric J85 turbojet engine, had a length of 23 feet and a wing span of 16 feet.

The compact HiMAT was an evolutionary concept, originating during the M2 lifting body program of the 1960s. To test a variety of lifting body shapes in flight, an innovative NASA engineer at Edwards built a twin-engine radio-controlled model that carried the smaller test models high into the sky and made 120 test drops. Typical remotely piloted vehicles (or RPVs) used an autopilot system and had restricted maneuverability. The Edwards aircraft, on the other hand, was completely controlled from the ground, using instrument references. By the late 1960s, Edwards personnel were flying an actual lifting body test configuration,the Hyper III, in drop tests from a helicopter. Veteran fliers who flew the model by remote control found it a remarkable experience. "I have never come out of a simulator emotionally and physically tired as is often the case after a test flight in a research aircraft," one pilot said. "I was emotionally and physically tired after a 3-minute flight of the Hyper III," he admitted. Although remote flight research continued, demands of the YF-12 Blackbird program and other projects kept it at a low level. Still, significant progress occurred. The Edwards team took a Piper Twin Comanche fitted with an electronic fly-by-wire system, added a television system for a remote pilot, and turned it into a successful remotely piloted aircraft from takeoff to landing. Although a backup pilot flew in the cockpit, the remote operators practiced stalls, stall recoveries, and even made precise instrument landing approaches. In the early 1970s, these skills were translated into an applicable test program to investigate stall and spin phenomena after several fighter planes were lost in spinning accidents. NASA let contracts to McDonnell Douglas for three 3/8-scale models of the F-15. Each model cost $250,000; a full-sized plane cost $6.8 million. Piloted from the ground and released from a B-52 at high altitude, the model F-15 program yielded useful information for final revisions of the operational Air Force fighter. The remote pilots doing the flying found the spin tests quite challenging: the heart beats of pilots in normal, manned flights went from 70-80 per minute to 130-140 during the remotely piloted drop tests.

The Highly Maneuverable Aircraft Technology test model (HiMAT), shown during a test flight. A modular design allowed engineers to test a variety of wings, control surfaces, and different structural materials.

The remotely controlled flight tests were controversial. Extensive ground support systems were nearly as expensive for remote flight operations as they were for manned aircraft. Still, remotely controlled flights were useful; models offered a cost-effective method for testing esoteric designs; they were obviously advantageous in dangerous flight maneuvers. The positive factors were convincing as NASA and the military services pondered exotic configurations and materials of combat planes for the 1990s and beyond. The logic for a test vehicle like the HiMAT was unusually strong.

The HiMAT structure itself was composed of various metal alloys, graphite composites, and glass fibers. It had sharply swept wings, winglets, and canard surfaces --considered aeronautically avant garde when the first plane flew in 1978. Carried aloft by a B-52, the HiMAT was remotely --and safely-- flown through a series of complex maneuvers at transonic speeds. The HiMAT was designed as a modular vehicle so that wings, control surfaces, and structural materials could be evaluated at a fraction of the cost of building a full-sized aircraft. The HiMAT's changing configurations suggested the possible shapes of aircraft to come.

While the HiMAT continued to test alternative design ideas, flight test specialists nonetheless recognized the persistent value of full-sized manned aircraft. The result was the Grumman X-29, a plane whose dramatic configuration matched that of the HiMAT. The X-29 had a single, vertical tail fin and canard surfaces --not unique in the 1980s. What made the X-29 so fascinating was its sharply forward-swept wings.

The forward-swept wing had precursors in German designs of World War II. In 1944, Junkers put such an experimental jet into the air--the JU-287. The war ended before extensive flight tests could be carried out, but the JU-287 quickly revealed one of the major problems of any swept forward design: structural divergence. Lift forces on wings cause them to bend slightly upward. When the wings sweep forward, this force tends to twist the leading edge upward, increasing lift and the bending motion until the wing fails. One solution was to keep the wing absolutely rigid, but conventional metal construction made such wings so heavy they were impractical. Although swept forward wings occasionally appeared on various aircraft in the postwar era, construction and weight problems proved intractable. The solution appeared in the form of composites, affording wings of light weight but high strength.

Grumman had submitted an unsuccessful HiMAT design, which ran into severe wing-root drag problems. A forward-swept wing seemed to offer answers, and the company had quietly pursued the idea. NASA also became interested, and the DoD eventually agreed to support a radical new design. NASA became responsible for technical support and flight testing. In 1987, the plane was officially announced as the X-29, the first new "X" aircraft developed by the United States in more than a decade. The fuselage took shape very quickly, since the forward section came from a Northrop F-5A. Landing gear came from the General Dynamics F-16A, and the engine was adapted from a General Electric power plant developed for the McDonnell Douglas F-18 Hornet. At first glance, the X-29 seemed a sorry aeronautical compromise, merely incorporating bits and pieces from other planes. But its wings and related design elements made it truly unique. Moreover, it was highly unstable.

When the X-29 made its first flight in 1984, the forward-swept wings and canard surfaces were its most distinguishing characteristics. In swept back wings, controllability became a problem as increasingly turbulent air flowed over the wing tips and tail surfaces. The X-29's wing tips, however, were always moving in comparatively undisturbed air, enhancing controllability at high speeds, and the canard surfaces also operated in an air stream much less turbulent than that around the tail. The rigid wing of the X-29 owed much to composites and the way they were layered in relation to the angle of the wing and aerodynamic stresses, overcoming the tendency to structural divergence.

With its swept forward wing and composite construction, the X-29 offered weight and drag reduction of as much as 20 percent compared to conventional design and fabrication methods.

Among the electronic advances of the X-29, the most fascinating related to its inherent instability. Most planes were built to be stable in flight, returning to straight and level flight if diverted. In a dog fight, such placidity could be fatal. The F-16 jet fighter was built to be about 5 percent unstable, but the X-29 was built to be about 35 percent unstable. This extreme instability was more than any pilot could manage, so a trio of flight computers were developed to keep the plane under control while allowing the pilot a remarkable latitude in terms of maneuverability. At a rate of 40 times per second, the computers analyze the plane's attitude and decide what is necessary to keep the plane under control while responding to the pilot's inputs. This allows for some unusual flight maneuvers which could contribute to more agile combat planes in the future. For one thing, the X-29 could "levitate" in flight --climbing while maintaining a straight and level attitude.

Exotic experimental military planes represented only one of several areas of NASA's study. During the 1970s, the general aviation sector became increasingly robust. Most Americans knew little about this remarkably diverse segment of American aviation, which included all aircraft except those flown by commercial airlines and the armed services. There were about 2400 scheduled airliners in service during the 1970s and 4300 in the 1980s, while the general aviation fleet climbed from 150,000 to 220,000 aircraft, ranging from propeller-driven single engine planes to multimillion dollar executive jets. Sales of general aviation aircraft represented a significant contribution to America's favorable balance of payments, since 90 percent of the world's fleet of general aviation types originated in American factories. Given the scope of general aviation operations in the United States and the significance of American domination of the world market for this sector, NASA's attention was probably overdue when the agency began comprehensive studies during the late 1970s. Results came very quickly as more than a dozen production and prototype designs incorporated features derived from relevant NASA studies.

Winglets were incorporated into the design of the Learjet model 55 (photo courtesy of Gates Learjet Corporation).

One distinctive hallmark of NASA's general aviation investigations was the wingtip winglet, a device to smooth out distorted air flow, resulting in improved wing efficiency and enhanced fuel economy. During the 1980s, a number of high performance business jets, such as the Learjet, as well as late-model transports built by Boeing and McDonnell Douglas used this innovation. The agency also developed a new high performance airfoil for general aviation; the GAW-1. A separate research effort went into stall/spin problems, using radio-controlled scale models as well as several different full-sized operational aircraft. There were additional programs to probe exhaust and engine noise, engine efficiency, and the use of composites. A special investigation of crash survivability tested the airframes of planes as well as injuries to passengers, represented by carefully instrumented anthropomorphic dummies. A huge drop tower let the test planes plunge onto a typical runway; test results were useful to many aviation industry firms, including manufacturers of aircraft seats, seat belts, and body restraint systems.

In Cooperation with the Federal Aviation Administration, crash tests were aimed at developing better protection for pilots and passengers in general aviation aircraft. Crash tower and damaged plane (inset).

Satellites and Space Science

During the 1970s, the number of American payloads put into space by rocket boosters diminished as mission planners waited for the shuttles to become operational. When the shuttles began flying with payloads in the 1980s, this did not mean that NASA's expendable rocket launches ceased. Several rocket launches had already been scheduled, and NASA also intended to maintain this capability as a backup through the mid-1980s. NASA boosters orbited a variety of communications and environmental satellites as well as several spacecraft involving space science. Moreover, the audacious Voyager continued its richly rewarding "grand tour" of the outer planets. Shuttle launches may have gotten the lion's share of news coverage, but rocketed payloads continued to demonstrate their share of utility and value in space exploration.

Meteorological satellites and other Earth-oriented space craft expanded their essential roles in contemporary society. During 1981, another Geostationary Operational Environmental Satellite (GOES-5) went into Earth-synchronous orbit. In addition to expanded hurricane observations in the Caribbean zone, GOES-5 tracked Gulf Stream currents for fishermen and others with marine interests, provided invaluable data for weather-casters, and warned citrus growers about potentially crop-killing frosts. The National Oceanic and Atmospheric Administration (NOAA) not only supplied vital data on ocean temperatures and wave patterns with NOAA-7; the multi-mission spacecraft conducted a variety of atmospheric and tidal measurements while monitoring solar particle radiation in space, alerting manned space missions and commercial aircraft of potentially hazardous conditions.

This network expanded with the launches of GOES-6 and NOAA-8 in 1983. The latter joined a space-based search and rescue system cooperatively operated by the United States, France, Canada, and the Soviet Union. Known as the Sarsat-Cospas network, the satellites of the participating countries could pinpoint the locations of emergency beacons aboard ships and aircraft in distress. Within a few months of its becoming operational, the rescue network had saved some 60 lives around the globe. Landsat-4, launched in 1982, experienced transmission failures, so Landsat-5 took over during 1984, continuing vital coverage for forestry, agriculture, mineral resources, and other uses. Also during the 1980s, NASA launched a series of new Intelsat communications satellites to replace older models in geosynchronous orbits above the Indian, Pacific, and Atlantic oceans.

Nonetheless, space science payloads and planetary probes continued to be the most dramatic performers. Following the encounter of Voyager 1 with Saturn in 1980, Voyager 2 made an even closer pass in the summer of 1981. These visits turned up considerable new information on Saturn's rings, moons, and weather systems, posing a number of new questions for planetary scientists. Continuing analysis of Pioneer Venus 1 also seemed to raise as many new issues as it closed. Launched in 1983, the Infrared Astronomical Satellite was a joint project of NASA and scientific centers in the Netherlands and Great Britain. During its 10-month lifetime, the international satellite detected new comets, analyzed infrared signals from a number of new galaxies, and yielded data that suggested many of them may be merging or colliding with each other.

Planetary probes continued to turn up surprising insights into the nature of our solar system. Four and a half years after uncovering a wealth of new data on Saturn and its spectacular rings, Voyager 2 approached Uranus in January 1986. By the time the intrepid Voyager completed its flyby, the spacecraft had revealed more information about the planet and its company of moons than observers had learned since its discovery by the English astronomer William Herschel over 200 years ago.

The spacecraft's arrival represented something of a tour de force for the JPL, managers of Voyager's aptly named "Grand Tour of the Solar System." JPL's navigators had to place the spacecraft within less than 200 miles of a point between the planet's innermost moon, Miranda, and the planet's rings. Having traveled 1.8 billion miles from Earth, Voyager 2 now whipped toward its goal at 50 times the speed of a pistol bullet. Commands from JPL to Voyager took 2 hours and 45 minutes to arrive. Unless the JPL crew did everything correctly, Voyager 2 might miss the gravitational sling from Uranus to send it on towards its rendezvous with Neptune in 1989. More important, engineers had to know the exact location of the Voyager so that its cameras would record something planetary instead of the infinite blackness of space. "That feat," explained a reporter for the National Geographic, "is equivalent to William Tell shooting an arrow in Los Angeles and hitting an apple in Manhattan." Many potential glitches were avoided, such as breaking into the onboard computer programs to fine tune the thrusters; commandeering another backup computer to improve the rate of image processing; and dispatching further signals to help Voyager perform in a colder, darker environment than was the case for its Saturn flyby.

There were other snarls as well, but Voyager 2 carried on superbly, turning up evidence of 10 new moons besides the five known orbs circling Uranus. Miranda, the smallest of the five, proved especially dramatic with a tortured surface that included an escarpment 10 times deeper than the Grand Canyon. The various moons represented a geological showcase, with mountains up to 12 miles high, plains dotted with craters, and sinuous valleys that may have been gouged out by glaciers. Voyager 2 also captured other curiosities about Uranus, including its offset magnetic field, fascinating ultraviolet sheen called an "electroglow," and erratic atmospheric patterns. Another mission to Uranus might be decades, or even centuries away. But the Voyager's legacy promised to give scientists and astronomers considerable data to ponder in the meantime.

Shuttle operations

At liftoff, the Shuttle looked and sounded like an oversized rocket booster with wings. Power for the launch came from a combination of propulsion systems. A pair of solid-fuel booster rockets straddled a huge propellant tank filled with liquid hydrogen and liquid oxygen; the Shuttle itself perched atop the cylindrical walls

104

of the propellant tank, which fed the trio of Space Shuttle main engines mounted in the Shuttle's tail. During the initial ascent phase, all five propulsion systems drove the Shuttle upwards. Following burn-out of the solid-fuel boosters, the empty casings separated from the external tank and parachuted back to Earth, where they were recovered from the ocean, refurbished, and packed again with segments of solid fuel. The Shuttle's liquid-hydrogen main engines continued to fire, drawing propellants from the external tank. When the tank was empty, it too was jettisoned and destroyed by intense heat during its descent through Earth's atmosphere. A pair of maneuvering engines plus batteries of small rocket thrusters on the Orbiter refined its orbital path as needed and provided maneuvering capability during the mission.

Compared to the Apollo spacecraft, the Orbiter was huge, with a length of 120 feet and a wingspan of 80 feet. As many as seven crew members could live and work in the flight deck area, and the cargo bay represented an additional payload or workspace area measuring 60 feet long by 15 feet in diameter. The Shuttle was designed to carry payloads of 65,000 pounds to orbit at an attitude of 230 miles (smaller payloads allowed orbits of up to 690 miles), return to Earth, and land with payloads of 32,000 pounds (such as a malfunctioning satellite). NASA contended that the ability to reuse the booster rocket casings and the ability of Orbiters to make repeated missions made the Space Shuttle an extremely cost-effective space vehicle for years to come. Because of all the tiles on the Orbiter, personnel associated with the program often joked about the "flying brickyard," but there was great enthusiasm about the Space Transportation System, or STS.

Although launches occurred at the Kennedy Space Center, and plans called for most Orbiter flights to finish there on a special landing strip three miles long, contingencies allowed for alternative landing sites at Vandenberg Air Force Base and Edwards Air Force Base in California, at White Sands, New Mexico, and at selected emergency runways around the world. In any case, the first few landings were planned for the broad expanses of the dry lake at Edwards; the Orbiter would be carried back to KSC from any remote site atop the specially modified Boeing 747 ferry aircraft. There were only five landings at Kennedy Space Center before a blown nose wheel tire at the end of the 16th (51-D) mission shifted all subsequent touchdowns to Edwards. Some earlier flights had been diverted from Kennedy because of weather; the Boeing 747 transporter definitely proved its value in returning Orbiters from Edwards, White Sands, and Vandenberg. Following the nose wheel incident, engineers planned changes for Orbiter landing gear as well as improvements to the Kennedy landing site.

Concerns about tiles and engines kept the first Orbiter for flight missions, the Columbia, grounded at KSC for nearly two years. In the meantime, other Shuttle crews kept their flying skills sharp by participating in further drop tests of the Enterprise and by training flights in a Grumman Gulfstream modified to imitate an Orbiter's landing characteristics. Crew members and trainees practiced experiments and other tasks in a microgravity environment through long training missions in a converted Boeing C-135 transport. These missions also tested theories about the nature of nausea ("motion sickness") caused by disorientation in space--a severe problem for crew members during long space missions. The plane would fly high, arching parabolas in the sky, giving trainees several seconds of "weightlessness" at the top of each stomach-churning climb. The training missions might last several hours-repeated climbs, nose-overs, and rapid descents before the next upward surge. For those aboard the plane, all this could be either highly exhilarating or very loathsome. Officially, NASA's C-135 was designated the Reduced Gravity Aircraft; unofficially, hapless trainees dubbed it the "vomit comet," "barf buzzard," and "weightless wonder."

Finally, long hours of flight training and grueling sessions in electronic simulators came to an end. The Columbia's flight crew, astronauts John Young and Robert Crippen, joked that they had spent so much additional time in the electronic simulators that they were "130 percent trained and ready to go." Their inaugural flight was set for 10 April 1981. But the Columbia mission, like others to follow, was scrubbed at the last minute on a technicality. Two days later, the countdown for Columbia matched a day of perfect weather at KSC, and the Space Shuttle thundered off into space, boosted by 7 million pounds of thrust from its solid-fuel rockets and liquid-hydrogen engines.

Reaching an altitude of 130 nautical miles, the Columbia's crew settled into orbit for a two-day mission. The Orbiter carried no cargo except an instrumentation package to record stresses during launch, flight, and landing, plus a variety of cameras. One of these, a remote television camera aboard the Orbiter, revealed gaps around the tail section, where some tiles apparently worked loose during launch. As the crew prepared for descent back to Earth, mission controllers were quietly concerned, worried that other tiles in critical areas along the Orbiter's underside might have fallen off as well. At a blinding speed of Mach 24, Columbia began its searing reentry back into Earth's upper atmosphere, where the intense heat of atmospheric friction built to over 3000° F. There were some anxious moments as the plummeting spacecraft became enveloped by a blanket of ionized gases that disrupted radio communications. At 188,000 feet, as the Columbia slowed to only Mach 10, mission control heard a welcome report from Crippen and Young that the Orbiter was performing as planned. A long, swooping descent and a series of planned maneuvers bled off excess speed and brought the spacecraft in over the Edwards area. Parked in cars, jeeps, and campers all around the edge of the landing area, an estimated 500,000 people had come to observe the Shuttle's return. The sharp crack of a sonic boom snapped across the desert, and the crowd soon saw the Columbia, now slowed to about 300 MPH, make its final descent and touchdown, a true, "spaceliner" symbolizing a new era in astronautical ventures.

For all its teething problems, the Shuttle performed remarkably well through five years and 24 successful missions. Inevitably, there was some fine tuning and reworking of numerous tiles before a second launch of Columbia in November, the first spacecraft to return to orbit. During 1982, three more missions marked the end of flight tests and the beginning of missions to deploy satellites. The next year, four additional missions included three in the new orbiter, Challenger, ending on Columbia's flight with the ESA's "Spacelab" aboard. There were six crew members, a record number for a single spacecraft, including Ulf Merbold, a German who represented the ESA. These flights in 1983, which counted America's first woman in space (Sally Ride) as well as the first black American (Guion Bluford), not only launched additional American and international payloads, but also significantly increased activities in space science, particularly with the Spacelab mission. To deploy satellites from the cargo bay, the crew relied on a unit called the Propulsion Assist Module, or PAM, introduced on the STS-5 mission in 1982. In the payload deployment sequence, the remote manipulator system lifted the satellite out of the Orbiter cargo bay. The Orbiter then maneuvered away; the PAM attached to the satellite automatically fired about 45 minutes later boosting the payload higher. The organization owning the satellite then took over, using thrusters on the satellite to circularize its orbit, checking out its systems, and making the satellite operational. Although the PAM booster was augmented by other systems, many payloads could be left in orbit after simply lifting them out of the cargo bay with the remote manipulator system.

The orbiter Discovery joined the fleet in 1984, and Atlantis followed in 1985. The demographics of the orbiter crews reflected growing diversity, encompassing more women, Canadians, Hispanics, Orientals, assorted Europeans, a Saudi prince, a Senator, E. J. "Jake" Garn, and a Congressman, Bill Nelson. The various missions engaged astronauts in extended extravehicular activity, such as untethered excursions using manned maneuvering units. In Mission STS-11 (41-C) in 1984, an astronaut using one of these units assisted in the first capture of a disabled satellite, the Solar Maximum payload (Solar Max), followed by its repair and redeployment. The mission also had the task of placing a new satellite in orbit. Scheduled for deployment was the Long Duration

Exposure Facility, a 12-sided polyhedron measuring 14 feet in diameter and 30 feet long. It carried several dozen removable trays to accommodate 57 experiments put together by some 200 researchers from eight countries. After being lifted out of the Challenger, the big structure was to stay in orbit for a year, awaiting its return on a different Shuttle flight.

For the crew aboard Challenger, the biggest task was the first planned repair of a spacecraft in orbit. The Challenger's thrusters boosted it 300 miles higher to intercept the Solar Max satellite. After some difficulties, due to the satellite's tumbling motion, it was finally stabilized and cranked down into the cargo bay by the remote manipulator system (RMS). After a night's rest, George Nelson and James van Hoften donned space suits and went to work on the balky satellite, replacing a faulty attitude control module and some electronic equipment for one of its instruments. Sent back into orbit, the Solar Max's repair job in space saved millions of dollars. Later the same year, during STS-14 (51-A), the crew of Discovery had to retrieve a pair of errant satellites placed in improper orbits by faulty thrusters. Although the Canadarm managed to capture the satellites, they would not drop into the cradles in the cargo bay for their return to Earth, and the mission specialists had to manhandle each one aboard before closing the cargo bay doors. These missions conclusively demonstrated the Shuttle's ability to recover, repair, and if necessary, refuel satellites in orbit. The DoD also made two classified missions in 1985.

Mission STS-22 (61-A), in October 1985, represented the fourth Spacelab flight and was notable for its eight-member crew-- requiring the eighth person to sleep aboard the Spacelab itself. Most significant was the special role of the West German Federal Aerospace Research Establishment, which managed the orbital work in which the Spacelab mission specialists carried out experiments in materials processing, communications, and microgravity. It was a highly successful mission, with only one memorable drawback. Aboard the Spacelab was a new holding pen for animals that contained two dozen rats and a pair of squirrel monkeys. The crew soon complained to controllers that the animal quarters needed modifications for any future flights. Food bars for the rats began to crumble, so that loose particles of rat food began floating around the Spacelab. Worse, some waste products from the rats also began to litter the Spacelab's atmosphere leading to pointed, scatological comments from the disgruntled crew.

Continuing missions carried a variety of American as well as international scientific experiments. One involved electrophoresis, in which an electric charge was used to separate biological materials; the goal in this case was the production of a medical hormone. Additional experiments emphasized vapor crystal growth, containerless processing, metallurgy, atmospheric physics, and space medicine, among other areas. The payload manifests for most missions were recognizably similar, listing satellites, experimental biomedical units, physics equipment, and so on. The manifest for STS-16 (51-D) in 1985 had a decidedly different quality, including a pair of satellites along with a "Snoopy" top, a windup car, magnetic marbles, a pop-over mouse named "Rat Stuff," and several other toys, including a yo-yo. For die-hard yo-yo buffs, a NASA brochure reported that the "flight model is a yellow Duncan Imperial." The news media gave considerable attention to the whimsical nature of the Toys in Space Mission, although the purpose was educational. The toy experiments were videotaped, with the astronauts demonstrating each toy and providing a brief narrative of scientific principles, including different behaviors in the space environment. The taped demonstrations became a favorite with educators--and the astronauts obviously delighted in this uncustomary mission assignment.

Despite occasional problems, Shuttle flights had apparently become routine--an assumption that dramatically changed with Challenger's mission on 28 January 1986.

On the morning of the flight, a cold front had moved through Florida, and the launch pad glistened with ice. It was still quite chilly when the crew settled into the Shuttle just after 8:00 A.M. Many news reports remarked on the crew's diversity; seven Americans who seemed to personify the nation's heterogenous mix of gender, race, ethnicity, and age. The media focused most of its attention on Christa McAuliffe, who taught social studies at a high school in New Hampshire. She was aboard not only as a teacher but as an "ordinary citizen," since Space Shuttle missions had seemed to become so dependable. Scheduled for a seven-day flight, the Challenger also carried a pair of satellites to be released in orbit.

NASA officials, leary of the icy state of the Shuttle and launch pad, waited two extra hours before giving permission for launch. When the Shuttle's three main engines ignited at 11:38 A.M., the temperature was still about 36° F, the coldest day ever for a Shuttle liftoff. After a few seconds, the solid-fuel boosters also ignited, and the Challenger thundered majestically upward. Everything appeared to be working well for about 73 seconds after liftoff. At 46,000 feet in a clear blue sky, the Shuttle was virtually invisible to exhilarated spectators at Cape Canaveral, but the telephoto equipment of television cameras captured every moment of the fiery explosion that destroyed the Challenger and snuffed out the lives of its crew. In the aftermath of the tragedy, stunned government and contractor personnel took action to recover remnants of the Shuttle and to begin a painstaking search for answers.

Answers were essential, because the three remaining Shuttles were grounded while the cause of the Challenger explosion was identified and corrected. Until that time, the United States could not put astronauts into space or launch any of the numerous satellites and military payloads designed only for deployment from the Shuttle cargo bay. Moreover, construction of the planned space station in Earth orbit relied entirely on the Shuttle's cargo capacity.

Detailed analysis of photography and Shuttle telemetry pointed to a joint on the right solid booster. It appeared that a spurt of flame from the joint (which joined fuel segments near the bottom of the booster) destroyed the strut attaching the booster to the bottom of the liquid hydrogen tank and burned through the tank itself. The tank erupted into a fireball, and the explosion blew apart the Challenger. Next, investigators had to understand the reasons for the faulty joint.

In the meantime, President Reagan appointed a special commission to conduct a formal inquiry--the Rogers Commission, named after its chairman, former Secretary of State William P. Rogers. The Rogers Commission discovered that NASA had been worried about the booster joints for several months. The specific problem involved O-rings, circular synthetic rubber inserts that sealed the joints against volatile gases as the rocket booster burned. It was believed that the O-rings lost their efficiency as boosters were reused; their efficiency was even less in cold weather. The Rogers Commission further discovered that NASA and managers from Thiokol, suppliers of the solid fuel boosters, had hotly debated the decision to launch during the night before Challenger's fatal flight.

The Rogers Commission report, released in the spring of 1986, included an unflattering assessment of NASA management, calling it "flawed," and recommended an overhaul to make sure managers from the Centers kept other top managers better informed. Other criticisms not only resulted in a careful redesign of the booster joints but also led to improvements in the Shuttle's main engines, a crew escape system, modified landing gear, alterations to the landing strip at Kennedy Space Center, and changes for a host of aspects in Shuttle operations. NASA originally planned to resume Shuttle flights in the spring of 1988, but nagging problems delayed new launches through the summer.

In the wake of Challenger's loss, other changes occurred. Some realignment would have occurred in any case, since NASA Administrator James Beggs, indicted for fraud and later completely exonerated, had vacated the position in December 1985. At the time of Challenger's loss, an interim leadership was in place; in the aftermath of Challenger, James C. Fletcher returned to NASA's helm again. But loss of the Shuttle colored many subsequent senior management reassignments in NASA, along with a reorganization of contractor personnel. Even though President Reagan authorized construction of a new Shuttle for operations by 1991, the existing fleet of three vehicles remained inactive for over a year and a half, severely disrupting the planned launch of civil and military payloads. For some scientific missions, desirable "launch windows" were simply lost, and other missions, rescheduled sometime in the future, were severely compromised in terms of scientific value. In the case of the Space Shuttle program, NASA had not only stumbled, but was left staggering.

Chapter 9

NEW DIRECTIONS (since 1986)

Although the flight of Voyager 2 past Uranus and on toward Jupiter represented a striking success, it was almost lost in the clamor triggered by the loss of Challenger. During the next several months, the agency's frustrations multiplied.

In 1986, Halley's Comet made its appearance again after an absence of 76 years. Halley was a valued astronomical performer. As the brightest comet that returned to the Sun on a predictable basis, scientists had adequate time to prepare for its reappearance. However, during Halley's dramatic swing across Earth's orbit, many American scientists lamented that no American spacecraft made a mission to meet it and make scientific measurements. Some U.S.launched satellites were able to make ultraviolet light observations, but only the ESA, Japan, and the Soviet Union had planned to send probes close enough to use cameras--ESA's Giotto probe came within 375 miles of Halley's nucleus. Critics charged that excessive NASA expenditures on the Shuttle had robbed America of resources to take advantage of unusual opportunities such as the passage of Halley's Comet.

In the aftermath of Challenger, NASA's hopes for recovery were further plagued by a rash of misfortunes. In May 1986, a Delta rocket carrying a weather satellite was destroyed in flight after a steering failure. One of NASA's Atlas-Centaur rockets, under contract to the U.S. Navy for the launch of a Fleet Satellite Communications Spacecraft, lifted off in March 1987, but broke up less than a minute later after being hit by lightning. During the assessment of the loss, a review board scolded NASA managers for making the launch into bad weather conditions that exceeded acceptable limits. In June, three rockets at NASA's Wallops Island facility were being readied for launch when a storm came in. Lightning hit the launch pad and triggered the ignition of all three rockets; frustrated engineers watched the trio shoot off in a hopeless flight over the Atlantic shoreline before crashing into the sea. In July, disaster hit NASA again when an industrial accident on the launch pad at Cape Canaveral destroyed an Atlas-Centaur upper stage on the launch pad, forcing cancellation of a military payload mission.

These embarrassments, and the brooding shadow of Challenger, dulled the otherwise bright successes. Early in 1987, determined launch crews had successfully put two important payloads into orbit. The GOES-7 environmental satellite went into operation, returning vital information of the formation of hurricanes in the Caribbean. An Indonesian communications satellite, Palapa B 2P, originally scheduled for a Shuttle launch, went into orbit aboard a Delta rocket launched from Cape Canaveral. While debate over the nation's space program persisted, NASA continued its space-work on several different projects. Taken collectively, they held considerable promise for many areas of both astronautics and aeronautics.

Artist's concept of the Hubble Telescope after deployment from the Orbiter. The most powerful telescope ever built, it is intended to allow scientists to look seven times farther into space than ever before. The ESA supplied the solar power arrays for this international project.

Astronautics

Resumption of Space Shuttle missions for which special payloads were developed may well trigger a renaissance in astronomical science, especially in the case of the Hubble Space Telescope. Weighing 12 1/2 tons and measuring 43 feet long, the Hubble Telescope with its 94.5-inch mirror is the largest scientific satellite built to date. All ground telescopes are handicapped by the Earth's atmosphere, which distorts and limits observations. The Hubble Telescope will permit scientists to collect far more data from a wide spectral range unobtainable through present instruments. The most alluring prospect of the Hubble Telescope's operation is the potential to search for clues of other solar systems and gather data about the origins of our own universe, perhaps solving once and for all the "big bang" theory of the universe as opposed to the steady state concept. Once in orbit, the telescope is expected to pick up objects 50 times fainter and 7 times farther away than any ground observatory; via electronic transmissions to Earth, the telescope can let humans see a part of the universe 500 times larger than has ever been seen before. A document issued by the JPL predicted that "primeval galaxies may be seen as they were formed, as they appeared shortly after the beginning of time." It is fitting that the Hubble Space Telescope is an international enterprise, with the ESA supplying the solar power arrays and certain scientific instruments as well as several scientists for the telescope's science working group.

Nor was the Hubble Space Telescope the only major effort in astronomy, astrophysics, or planetary research. NASA planned a new family of orbiting observatories, often developed with foreign partners, to probe more deeply into the background of gamma rays, infrared emissions, celestial x-ray sources, ultraviolet radiation, and a catalog of other perplexing subjects. There were also several bold planetary voyages to be launched. In collaboration with the Federal Republic of Germany, the Galileo mission to Jupiter (requiring a six-year flight

112

after launch from the Space Shuttle) called for an atmospheric probe to be parachuted into the jovian atmosphere while the main spacecraft went into orbit as a long-term planetary observatory. The Magellan mission envisioned a detailed map of the planet Venus; Ulysses (planned with ESA) was designed to explore virtually unchartered solar regions by flying around the poles of the Sun. All of these missions were targeted for the late 1980s and early 1990s; creative scientists and engineers were also concocting ambitious projects for the twenty-first century.

During 1986 and 1987, Sally Ride, of NASA's astronaut corps, spearheaded a special NASA Headquarters task force charged with determining new priorities for the nation's space program. The task force eventually narrowed its recommendations to four principal possibilities. The first concerned Earth studies to gain knowledge for protection of the world's environment. A second proposal focused on accelerated robotic programs to explore the Moon and other bodies in the solar system. These two areas of activity were already implicit in many NASA programs underway or planned for the near future. The final two proposals were particularly exhilarating to partisans of manned exploration, since they projected a permanent human outpost on the Moon and subsequent manned expeditions to Mars. In the spring of 1987, NASA made a determined step towards lunar and martian missions by creating the Office of Exploration to begin planning for these programs. NASA's plans for an operational space station, while not crucial for these goals, were nevertheless important, since the station could play a major role in their support.

The first technically reasoned studies of a space station began in the late 1930s, when Arthur Clarke and his friends in the British Interplanetary Society began publishing proposed designs. Rocketry in World War II seemed to make these speculations far less sensational to the postwar generation. In March 1952, the popular American magazine, Colliers, startled some readers but fascinated others with a special edition on space exploration. One of the more dramatic articles featured a space station shaped like a huge wheel, 250 feet in diameter, designed to rotate in order to provide artificial gravity for the station's inhabitants.

During the next three decades, variations of the Colliers design and other space station structures appeared in a variety of popular and technical journals. Some early ideas, like the need for artificial gravity, persisted for a long time before finally disappearing (except for special requirements like centrifuge experiments). Others, like modular structures, free-flying "taxis," and a stationary facility for zero-gravity activities remained staples of space station thinking. With the organization of NASA in 1958, space station planning took on a more practical aspect as part of a national commitment to space exploration. Within two years of its founding, NASA had organized a committee within the Langley Research Center to study technology required for space stations.

The process of deciding the design of a space station and its uses consumed over two decades and several million dollars. A significant milestone occurred in January 1984, when President Ronald Reagan endorsed the Space Station Freedom program in his State of the Union message. Meanwhile, NASA and contractor space station studies proceeded through several variations before one design was designated by NASA as the "baseline configuration." This structure, which emerged during 1987-88, was scaled down in size because of budgetary constraints and the reduced number of Shuttle flights after the loss of the Challenger. A primary concern was to put a station in operation by the mid-1990s. At the same time, NASA publicized what it called a phased approach, giving the agency an option for adding several large components once the basic space station was in place. The revised baseline configuration called for a horizontal boom about 360 feet long, with pairs of solar panels at each end to generate 75 kilowatts of power. At the center of the boom, four pressurized modules, linked together, provided the focus of manned operations in a 220-mile orbit above the Earth. The American space station initiative included an invitation to foreign partners to share in its planning and operation; refining the details of this partnership engaged negotiators from the United States, Canada, Japan, and the ESA over the next four years. The toughest negotiations involved ESA. The Europeans wanted to insure free access to the space station and to guarantee some technology transfer in return for their contributions to station development. The foreign partners also strenuously resisted plans for significant space station activities by the American armed services. The United States and its international partners agreed to limit space station uses to "peaceful purposes," as determined by each partner for its own space station module. The final documents were signed by ESA, Japan, and Canada in September 1988. The United States was responsible for a laboratory module and a habitation module for the crew. The Europeans and Japanese were each responsible for the two additional laboratory/experimental modules; Canada was to supply a series of mobile telerobotic arms for servicing the station and handling experimental packages. Plans called for eventual use of manned and unmanned free-flying platforms for special missions away from the station. Eventually, the station might add solar-dynamic power generators and two vertical spines, located on either side of the module cluster and joined by upper and lower booms, providing additional attachment points for external scientific equipment.

One of the many studies for the Space Station. Solar power panels at each end of the elongated truss structure supply electricity for the cluster of living and science modules at the center. Lab work will emphasize microgravity experiments in pharmaceutical research, development of flawless crystals for advanced supercomputers, and life sciences investigations such as the study of the behavior of living cells.

Aeronautics

Aeronautical research proceeded along several lines. The Grumman X-29 began flying additional missions to test upgraded instrumentation systems. With Air Force cooperation, a considerably modified F-111 carried out flight tests using a Mission Adaptive Wing, in which the wing camber (the curve of the airfoil) automatically changed to permit maximum aerodynamic efficiency. With the DoD, NASA launched development of a hypersonic aircraft, the X-30, tagged with the inevitable acronym: NASP, for National Aero-Space Plane. Plans called for a hydrogen-fueled aircraft that would take off and land under its own power. The plane would streak aloft at Mach 25, and be able to operate in a low Earth orbit much like the Shuttle, or cruise within the Earth's atmosphere at hypersonic speeds of Mach 12. Its ability to sprint from America to Asia in about three hours encouraged the news media to refer to it as the "Orient Express." A series of developmental contracts awarded during 1986 and 1987 focused on propulsion systems and certain aircraft components; an experimental, interim test plane was several years away.

One version of the NASA-developed propfan, mounted on a production airliner for flight tests.

Other flight research represented a totally different regime of lower speeds and emphasis on fuel efficiency. Even though jet fuel prices dropped in the mid-1980s, the cost was still five times the amount in 1972, and represented a significant percentage of operating costs for airlines. For that reason, airlines and transport manufacturers alike took an intense interest in a new family of propfan engines sparked by NASA's earlier Aircraft Energy Efficiency Program. Using a gas turbine, the new engine featured large external fan blades that were swept and shaped so that their tips could achieve supersonic velocity. This would allow the propfan to drive airliners at jet-like speeds, but achieve fuel savings of up to 30 percent. Different trial versions of multi-bladed propfan systems were in flight test beginning in 1986, with operational use projected by the early 1990s.

Investigation of rotary wing aircraft continued, even as the experimental XV-15 tiltrotor craft evolved into the larger V-22 Osprey, built by Boeing Vertol and Bell Helicopter for the armed services. A joint program linked the United Kingdom, NASA, and the DoD for investigation of advanced short-takeoff and vertical- landing aircraft. Based on the sort of concept used in the British Harrier "jumpjet" fighter, designers began wind tunnel tests of aircraft that could fly at supersonic speed while retaining the Harrier's renowned agility.

Several new NASA facilities promised to make significant contributions to these and other futuristic NASA research programs. NASA's Numerical Aerodynamic Simulation Facility, located at Ames and declared operational in 1987, relied on a scheme of building-block supercomputers capable of one billion calculations per second. For the first time, designers could routinely simulate the three-dimensional airflow patterns around an aircraft and its propulsion system. The computer facility permitted greater accuracy and reliability in aircraft design, reducing the high costs related to extensive wind tunnel testing. At Langley, a new National Transonic Facility permitted engineers to test models in a pressurized tunnel in which air was replaced by the flow of supercooled nitrogen. As the nitrogen vaporized into gas in the tunnel, it provided a medium more dense and viscous than air, offsetting scaling inaccuracies of smaller models--usually with wing spans of three to five feet-- tested in the tunnel.

Nonetheless, large tunnel models and full-sized aircraft still provided critical information through wind tunnel testing. For years, the world's largest tunnel was a 40 x 80-foot closed circuit tunnel located at Ames. It was a low speed tunnel (about 230 MPH), but its size permitted tests of comparatively large scale models of aircraft. As Ames became more involved in tests of helicopters and new generations of V/STOL aircraft, the need for a full-size, low speed tunnel became more apparent. The result was a new tunnel section, built at an angle to the existing 40 x 80-foot structure. Completed in 1987, the addition boasted truly monumental dimensions, with a test section 80 feet high and 120 feet wide, three times as large in cross-section as the parent tunnel. Overall, the new structure was 600 feet wide and 130 feet high. The original tunnel's fans were replaced with six units that increased available power by four times and raised the speed of the original tunnel from 230 to 345 MPH.

The new addition, with a speed of 115 MPH, was an open-circuit tunnel, using one leg of the original tunnel as the air was drawn through the bank of six fans. The very large cross-section of the 80 x 120 tunnel minimized tunnel wall boundary effects, which could seriously distort tests of full-sized helicopters and V/STOL aircraft. Although the tunnels could not be run simultaneously, technicians could set up one test section while the other was in operation.

Spinoff

NASA had evolved into an agency of a myriad activities. During the peak of Apollo program research in the 1960s, NASA became committed to the "spinoff" concept--space technology and techniques with other applications. A series of organizational efforts to publicize and encourage practical application of new technologies had been consistent ever since. The Apollo era's legacy included considerable biomedical information and physiological monitoring systems, developed for manned space flight, that enjoyed widespread implementation in hospitals and medical practice generally. In other areas, development of the Saturn launch vehicles prompted widespread improvements in bonding and handling exotic alloys, cryogenic applications, and production engineering.

The energy crunch of the 1970s prompted NASA to consider ways of transferring its considerable expertise in insulation materials, solar energy, heat transfer, and similar topics to the market place. In the process of analyzing a completely different problem, an investigation into the problems of hydroplaning (the tendency of aircraft tires to skid on wet runways) resulted in the technique of grooving runway surfaces. Similar treatment of high-speed highways was an obvious application; all this led to something called the International Grooving and Grinding Association, a conglomeration of some 30 obviously specialized companies in America, Europe, Japan, and Australia. Such an association might sound amusing, but their treatment of airports, highways, sidewalks, warehouse floors, and industrial sites has demonstrably enhanced industrial and human safety.

In the energy-conscious era of the 1970s, NASA's operational experience found many new applications. This prototype water heater, warmed by solar cells, was installed on a home in Idaho.

In a different context, NASA developed an entity called the Computer Software Management and Information Center, known by a singularly impressive acronym, COSMIC. Managed by the University of Georgia, COSMIC represented over 1400 NASA computer programs that were either directly applicable to customer needs or might be modified for specific requirements. The COSMIC library had provided answers for structural analysis as well as vehicular design; developed layouts for complex electronic circuitry; assisted architects in assessing energy requirements and reducing plant noise, and so on. Patrons of COSMIC thus saved invaluable time and millions of dollars by using available programs rather than developing a new one or risking serious design flaws by doing without.

A computerized structural analusis program perfected by NASA was used in the development of the Beechcraft Super King Air business plane.

These and other programs represented a significant NASA contribution to economic and commercial development. The "commercialization of space," a theme of President Ronald Reagan's space policy in the late 1980s, promised many more benefits stemming from renewed Shuttle missions and an operational space station. Advantages in metallurgy, biology, and medicine seemed the likeliest to be realized in the near future. These programs implied more and more reliance on manned flight, a situation that continued to disturb the practitioners of space science, underscoring a dichotomy in the nation's program that has persisted for many years.

In 1980, NASA's budget stood at $5 billion, and rose to $10.7 billion for the 1989 fiscal year. Manned space flight accounted for over half of that budget, while space science accounted for $1.9 billion, or about 18 percent. This share of funding for space science reflected a consistent pattern over the years, averaging about 20 cents of each NASA dollar. Critics of the space program often cited this difference in funding, and grumbled that so many Shuttle flights were scheduled for military missions. This fact, coupled with the need of 20 or more Shuttle missions to deliver space station components into orbit, meant fewer potential space science payloads. Critics also pointed out that the cost per pound of Shuttle missions exceeded early projections by a considerable margin, undercutting the original arguments in favor of the manned launch system. The Air Force had already, in the early 1980s, begun development of a family of expendable launchers, to reduce costs and provide alternatives to the possibility of a grounded Shuttle fleet. Many foreign customers found it economical to rely on the Ariane launch vehicle, operated under the authority of the ESA. NASA itself planned to use a new series of expendable launch vehicles to complement the Shuttle. Complicating the picture was the potential competition from a new Soviet shuttle vehicle, while ESA also had plans for a similar reusable spacecraft. Finally, the U.S. space commercialization policy prompted several U.S. companies to plan a variety of privately designed and built launch vehicles, which would also compete with NASA's own rocket launchers and the Space Shuttle.

With the launch of STS-26 (29 September 1988), the Space Shuttle Discovery marked NASA's first manned mission since the loss of Challenger and its crew two years earlier.

In 1990, the 75th anniversary of its founding as the NACA, the National Aeronautics and Space Administration, is a robust and diverse agency, experiencing continuing challenges in a diversified environment of air and space that it has helped to create.

SUMMARY

During the halcyon era between World War I and World War II, the NACA's work on airfoils, engine cowlings, icing, and other problems drew the attention of aeronautical engineers around the world. There were also institutional changes, especially in the 1930s, when the agency became more attuned to industry trends and became more politically aware in its interaction with congressional committees. World War II brought the most dramatic changes: research geared to national security; growth from one small facility to three spacious centers sited coast-to-coast; and ballooning budgets and personnel rosters. For all its successes, the agency also lost some of its luster as European advances in gas turbines and high-speed flight received postwar attention.

The postwar era entailed Cold War tensions and national security budgets that promoted advanced flight research. The NACA flourished. Cooperative programs with the military brought the X-1 and X-15 into being. These programs also moved the NACA out of the tradition of research and flight testing by adding responsibilities for design and program management as well. The old "advisory" committee had become a major R&D bureaucracy.

The shock of the successful Soviet launch of Sputnik in 1957 altered the NACA forever. Granted billion-dollar budgets by Congress, the new NASA was thrust into an international spotlight as America's answer to the Soviet Union for leadership in space exploration. With four new Centers, NASA rapidly developed skills in the novel field of astronautics. Personnel also had to build new skills as managers of huge budgets and mature aerospace contractors scattered across the continent. The spotlight of the space race also intensified the agency's problems when projects missed deadlines and when astronauts died. Still, Apollo was a successful effort and an historic achievement. While issues of American and Soviet competition for global influence colored the origins of the program and the triumphant voyage of Apollo 11, the new awareness of the fragile existence of Earth within our universe also fostered a promising spirit of international cooperation.

The post-Apollo era was not necessarily clear in terms of missions and purpose. The sense of urgency that spurred Apollo had dissipated. In aeronautics, NASA made sure progress in hypersonic flight and began highly beneficial programs to control pollution, reduce engine noise, and enhance fuel economy--programs that assumed growing importance in an environmentally conscious society. In astronautics, the Space Shuttle was a fascinating program, although critics maintained that it was a complex system with no major or scientific mission to justify its expense. A proposed Space Station, which would absorb numerous Shuttle flights, was plagued by budget issues; it was not expected to be operational until some time in the 1990s.

Meanwhile, the loss of Challenger in 1986 underscored the risk of relying so heavily on the Shuttle at the expense of expendable launch vehicles. Reorganizing priorities for military and civil payloads proved to be a frustrating exercise. A renewed wave of criticism concerning lower budgets for space science surfaced, a reminder of controversies over manned versus unmanned flights that had been going on since the early days of the space program. There was also concern stemming from various studies that noted the constraining effects that seemed endemic to large bureaucracies, as well as the demographic realities of a work force--heavily recruited in the 1960s--that might lose its sense of adventure as the time for retirement loomed.

In 1990, the 75th anniversary of its origins as the National Advisory Committee for Aeronautics, NASA nonetheless appears to be on a steady course. With new initiatives in commercial space programs and a broad spectrum of projects for applied science and technology in daily life, NASA surely has ventured far from its aeronautical origins in 1915. But the dynamics of flight--whether spacecraft or aircraft--still pervade the agency's principal activities. Beginning in 1988 with the STS-26 mission of the Discovery, manned missions aboard the Shuttle have resumed. At the same time, use of expendable launch vehicles have picked up, evidence that NASA planners are serious in attempting to broaden their options for getting payloads into orbit. Looking ahead, the Hubble Space Telescope is only one of many promising ventures in the area of space science and applications. The final agreements for international development of the Space Station have been signed. A broad spectrum of international scientific investigations are underway. NASA has also joined with the U.S. DoD and the United Kingdom pioneers in vertical takeoff and landing aircraft like the Harrier to foster the research and technology for an advanced short takeoff and landing aircraft, continuing a European connection that dates back to the founding of the agency in 1915. The forward swept wing X-29 continues an impressive flight research program; elsewhere, the development of low-speed propfan technology promises significant gains in fuel efficiency for subsonic airliners of the future.

The dynamics of flight promise to be just as challenging and fascinating in the future as they have been in the past.

BIBLIOGRAPHIC ESSAY

Background

An up-to-date aerospace bibliography prepared by the staff of the National Air and Space Museum not only provides an annotated, comprehensive guide to both American and international sources but also includes a fine review of other bibliographies: Dominick A. Pisano and Cathleen S. Lewis, eds., *Air and Space History: An Annotated Bibliography* (New York: Garland, 1988). For general coverage of flight, with emphasis on the years through World War I, see Charles H. Gibbs-Smith, *Aviation: An Historical Survey from Its Origins to the End of World War II* (London: Her Majesty's Stationery Office, 1970, Rev. 1985). The American chapter of early aeronautics is definitively recounted by Tom D. Crouch, *A Dream of Wings: Americans and the Airplane, 18751905* (New York: Norton, 1981). Joseph J. Corn, *The Winged Gospel: America's Romance with Aviation, 19001950* (New York: Oxford University Press, 1983), offers a thoughtful, interpretive analysis. For a combined survey of American aviation and space exploration, see Roger E. Bilstein, *Flight in America: From the Wrights to the Astronauts* (Baltimore: Johns Hopkins University Press, 19841987). A popular and useful survey of astronautics, with numerous illustrations, is Wernher von Braun and Fred I. Ordway III, *History of Rocketry and Space Travel* (New York: Thomas Y. Crowell, 1975). A series of scholarly essays with special attention to American topics is included in Eugene Emme, ed., *The History of Rocket Technology: Essays on Research, Development, and Utility* (Detroit: Wayne State University Press, 1964). The Pulitzer prizewinning study by Walter McDougal, *The Heavens and the Earth: A Political History of the Space Age* (New York: Basic Books, 1985), analyzes the American and Soviet space programs as part of the Cold War and technocratic trends. The NASA History Office has sponsored a series of monographs on air and space, most of which are noted below. A complete list of NASA History Series titles appears at the end of this book.

NACA and Aviation to 1958

The NACA's origins, technical contributions, and political evolution have been thoroughly assessed by Alex Roland, *Model Research: The National Advisory Committee for Aeronautics, 19151958*, 2 vols. (Washington, D.C.: U.S. Government Printing Office, 1985). The first volume represents an historical narrative; volume two contains annotated documentation. Roland criticizes the politicization of the agency. James R. Hansen, *Engineer in Charge: A History of the Langley Aeronautical Laboratory, 19171958* (Washington, D.C.: U.S. Government Printing Office, 1987), while covering the same time span, focuses on Langley's research functions. Both of these studies have strongly influenced this latest revision of *Orders of Magnitude. A history of the Lewis Research Center*, by Dr. Virginia Dawson, is in progress. The organization and early years of NACA's Ames facility are the subjects of Dr. Elizabeth A. Muenger, *Searching the Horizon: A History of Ames Research Center, 19401976* (Washington, D.C.: U.S. Government Printing Office, 1985).

General trends in the aviation industry can be traced in John B. Rae, Climb to Greatness: The American Aircraft Industry, 19201960 (Cambridge, Mass: MIT Press, 1968). For specific technical development by individuals and organizations in addition to the NACA, see Ronald Miller and David Sawers, The Technical Development of Modern Aviation (New York: Praeger, 1970). The fascinating story of the jet engine, and Europe's leadership in this field, can be found in Edward W. Constant II, The Origins of the Turbojet Revolution (Baltimore: Johns Hopkins University Press, 1980). The monographs by Roland and Hansen, cited above, represent other carefully argued viewpoints.

For an informative look at early rocket societies in America as well as abroad, see Frank H. Winter, Prelude to the Space Age: The Rocket Societies, 192440 (Washington, D.C.: Smithsonian Institution Press, 1983). On the background of German rocketry and Wernher von Braun, see the popularly written study by Frederick I. Ordway III and Mitchell R. Sharpe, The Rocket Team (New York: Thomas Y. Crowell, 1979), based on extensive interviews.

For a summary of the NACA's early postwar aerodynamic activities, see Hansen, Engineer in Charge. The story of the X-1 and the early challenge of the "sonic barrier" are detailed in Richard P. Hallion, Supersonic Flight: Breaking the Sound Barrier and Beyond (New York: Macmillan, 1972). There are further details in Hallion, On the Frontier: Flight Research at Dryden, 1946-1981 (Washington, D.C.: U.S. Government Printing Office, 1984). The story of Michael Gluhareff and the swept wing is recounted in an article by the same author, "Lippisch, Gluhareff, and Jones: The Emergence of the Delta Planform and the Origins of the Sweptwing in the United States," Aerospace Historian, 26 (March 1979): 110.

Origins of NASA through 1969

A series of NASA-sponsored histories covers the transition of the NACA to the new NASA and the progress of the Apollo program. The background of the IGY and America's initial plans to launch a satellite are the subject of Constance Green and Milton Lomask, *Vanguard: A History* (Washington, D.C.: Smithsonian Institution Press, 1971). The dramatic transition from the NACA to NASA and the difficulties of launching a coherent space program are clearly set out in Lloyd S. Swenson, Jr., James M. Grimwood, and Charles C. Alexander, *This New Ocean: A History of Project Mercury* (Washington, D.C.: U.S. Government Printing Office, 1966). Robert L. Rosholt, *An Administrative History of NASA, 19581963* (Washington, D.C.: U.S. Government Printing Office, 1966), details the bureaucratic organization.

As attention began to focus on possible lunar missions, politics and technology played interacting roles, a story that is set out by John M. Logsdon, The Decision to Go to the Moon: Project Apollo and the National Interest (Cambridge, Mass.: MIT Press, 1970). Unmanned exploratory missions to the lunar surface are the subject of Cargill Hall, Lunar Impact: A History of Project Ranger (Washington, D.C.: U.S. Government Printing Office, 1977).

Manned launches unquestionably provided drama during the space missions of the 1960s. The Mercury program is covered by Swenson, et al., in This New Ocean. For the official history of the next manned phase, see Barton C. Hacker and James M. Grimwood, On the Shoulders of Titans: A History of Project Gemini (Washington, D.C.: U.S. Government Printing Office, 1977). The Apollo missions (through Apollo 11), which formed the centerpiece of America's manned space effort during the decade, are the subject of Courtney G. Brooks, James M. Grimwood, and Lloyd S. Swenson, Jr., Chariots for Apollo: A History of Manned Lunar Spacecraft (Washington, D.C.: U.S. Government Printing Office, 1979). The success of Apollo required development of a family of large launch vehicles and a sophisticated launch complex. These topics are covered in Roger E. Bilstein, Stages to Saturn: A Technological History of the Apollo/Saturn Launch Vehicles (Washington, D.C.: U.S. Government Printing Office, 1980), and Charles D. Benson and William B. Faherty,Moonport: A History of Apollo Launch Facilities and Operations (Washington, D.C.: U.S. Government Printing Office, 1978).

Although launches from Cape Canaveral inevitably drew hundreds of thousands of enthusiastic spectators, public support of the space program was far from unanimous. A number of writers criticized the program as a cynical mix of public relations and profit- seeking; a massive drain of tax funds away from serious domestic ills of the decade; a technological high card in international tensions during the Cold War. See, for example, Edwin Diamond, The Rise and Fall of the Space Age (Garden City, N.Y.: Doubleday, 1964); Amitai Etzioni, The Moondoggle: Domestic and International Implications of the Space Race (Garden City N.Y.: Doubleday, 1964); Vernon van Dyke, Pride and Power; the Rationale of the Space Program (Urbana, Ill.: University of Illinois Press, 1964).

On the other hand, Richard S. Lewis, a highly regarded scientific journalist, has written a balanced assessment, The Voyages of Apollo: The Exploration of the Moon (New York: Quadrangle, 1974). Tom Wolfe, The Right Stuff (New York: Farrar, Straus and Giroux, 1979), is a scintillating essay that emphasizes personalities of the astronauts. Although astronauts are not necessarily considered skillful authors, Michael Collins, Carrying the Fire: An Astronaut's Journeys (New York: Farrar, Straus and Giroux, 1974), is an exceptionally well written memoir that is notable for its lucidity, as well as its modesty.

The Post-Apollo Years: 19691980

A trio of NASA-sponsored monographs deal with the principal programs of the early post-Apollo era. Edward C. Ezell and Linda Neuman Ezell, The Partnership: A History of the Apollo-Soyuz Test Project(Washington, D.C.: U.S. Government Printing Office, 1978), is a fascinating record of the negotiations and technical adjustments necessary to bring American and Soviet manned spacecraft together in orbit.

There had been considerable criticism of NASA's emphasis on manned missions, a bias that many observers felt had hindered progress in space science. This issue was somewhat ameliorated by the spectacular unmanned Mars probes of the late 1970s. The Ezell writing team detailed these activities in On Mars: Exploration of the Red Planet, 19581978 (Washington, D.C.: U.S. Government Printing Office, 1984.)

There was also a significant volume of space science undertaken in the manned missions of Skylab, carefully and skillfully explained by W. David Compton and Charles D. Benson, Living and Working in Space: A History of Skylab (Washington, D.C.: U.S. Government Printing Office, 1983).

Science is also an important theme in Clayton R. Koppes, JPL and the American Space Program (New Haven: Yale University Press, 1982), a book that also elucidates relationships between NASA and its contractors, including the academic community. Space science is the principal theme of Homer E. Newell, Beyond the Atmosphere: Early Years of Space Science (Washington, D.C.: National Aeronautics and Space Administration, 1980). As a central figure during the years of Vanguard through Shuttle plans of the early 1970s, Newell's is a valuable memoir. For a recent survey, see Paul A. Hanle and V. Chamberlin, eds., Space Science Comes of Age: Perspectives in the History of the Space Sciences (Washington, D.C.: Smithsonian Institution Press, 1982).

Elizabeth A. Muenger, Searching the Horizon: A History of Ames Research Center, 19401976 (Washington, D.C.: U.S. Government Printing Office, 1985) discusses this center's important role in aeronautics as well as astronautics.

NASA's continuing work in high-speed flight research is chronicled by Richard P. Hallion, On the Frontier: Flight Research at Dryden, 19461981 (Washington, D.C.: U.S. Government Printing Office, 1984,), a book that covers the X-15, lifting bodies, and the evolution of the Space Shuttle.

Jay Miller, The X-Planes: X-1 to X-29 (St. Croix, Minn.: Specialty Press, 1983), is a useful, heavily illustrated reference work. David A. Anderton, Sixty Years of Aeronautical Research, 19171977 (Washington, D.C.: U.S. Government Printing Office, 1978), is a concise, well illustrated summary. Although it focuses on Langley and offers little interpretation, it is a useful guide to NACA and NASA aviation programs.

NASA in the Shuttle Era

The NASA History Office is sponsoring a number of projects on various aspects of the Space Shuttle, planetary probes, applications satellites, space science, the space station, university/contractor relations, cultural responses to flight, and so on. While certain elements of these studies have been shared by the authors at professional meetings and NASA colloquia, publication of finished products is still pending. In the meantime a variety of NASA publications and other scattered sources can be consulted.

Hallion, On the Frontier , provides an informative survey of high-speed aeronautical experimentation as well as useful flight test information about the Shuttle. Howard Allaway, "The Space Shuttle at Work," NASA SP432 (1980), a NASA brochure released on the eve of Shuttle operational flights, nonetheless provides good technical background and mission plans.

The destruction of the Challenger is officially assessed in "Report of the President's Commission on the Space Shuttle Challenger" (Washington, D.C.: U.S. Government Printing Office, 1968), and offers insights into NASA's political, technical, and managerial characteristics. The agency became the target of many critical books and articles that not only dissected the Challenger incident but discussed perceived flaws throughout the NASA structure. See, for example, Joseph J. Trento, Prescription for Disaster: From the Glory of Apollo to the Betrayal of the Shuttle (New York: Crown Publishers, 1987). Alex Roland, "The Shuttle: Triumph or Turkey?" Discover, 6 (November 1985): 29-49, a cautionary assessment of the Shuttle, appeared three months before Challenger's last mission.

A sense of NASA's varied efforts in energy research, aeronautics, and space science over the past several years can be found in "NASA the First 25 Years, 19581983," NASA EP-182 (1983). NASA has released numerous brochures pertaining to specific projects and missions. See, for example, "Galileo to Jupiter: Probing the Planet and Its Moons," Jet Propulsion Laboratory, JPL 40015 (1979); Joseph J. McRoberts, "Space Telescope," NASA EP-166 (n.d.). These and a wide range of NASA news releases are well illustrated and useful sources. See also NASA's colorful and informative annual report, Spinoff (1976 to date), which includes programs that either are being applied or may be put to use.